AF735580

Environmental Research Advances

Marine Science and Technology

nova
science publishers

www.novapublishers.com

Environmental Research Advances

Activated Charcoal for Environmental Sustainability
Dipika Jaspal, PhD (Editor)
Arti Malviya, PhD (Editor)
Shraddha Sharma, PhD (Editor)
2023. ISBN: 979-8-88697-916-9 (Hardcover)
2023. ISBN: 979-8-88697-966-4 (eBook)

The Quantum Biosemiosphere
Jane Gray Morrison (Author)
Michael Charles Tobias, PhD (Author)
2021. ISBN: 978-1-68507-366-4 (Hardcover)
2021. ISBN: 978-1-68507-483-8 (eBook)

More information about this series can be found at
https://novapublishers.com/product-category/series/environmental-research-advances/

Marine Science and Technology

Numerical Modeling, Prediction of Ship Maneuvering and Hydrodynamics during Inland Waterway Transport
Peng Du (Editor)
Haibao Hu (Author)
Ying Wang (Author)
A. Ouahsine (Author)
Xiaopeng Chen (Author)
2022. ISBN: 978-1-68507-278-0 (Softcover)
2022. ISBN: 978-1-68507-575-0 (eBook)

Cetaceans: Evolution, Behavior and Conservation
Daniel Patón Domínguez (Editor)
2019. ISBN: 978-1-53614-998-2 (Softcover)
2019. ISBN: 978-1-53615-156-5 (eBook)

More information about this series can be found at
https://novapublishers.com/product-category/series/marine-science-and-technology/

Tan Min Pau
Mazlan Abd. Ghaffar
Muhd Danish-Daniel Abdullah
and Mohamad Nor Azra
Editors

Climate Change and Environmental Perturbations

Impacts on Biodiversity

nova
science publishers
www.novapublishers.com

NOTICE TO THE READER

The Publisher has taken reasonable care in the preparation of this book but makes no expressed or implied warranty of any kind and assumes no responsibility for any errors or omissions. No liability is assumed for incidental or consequential damages in connection with or arising out of information contained in this book. The Publisher shall not be liable for any special, consequential, or exemplary damages resulting, in whole or in part, from the readers' use of, or reliance upon, this material. Any parts of this book based on government reports are so indicated and copyright is claimed for those parts to the extent applicable to compilations of such works.

Independent verification should be sought for any data, advice or recommendations contained in this book. In addition, no responsibility is assumed by the Publisher for any injury and/or damage to persons or property arising from any methods, products, instructions, ideas or otherwise contained in this publication.

This publication is designed to provide accurate and authoritative information with regards to the subject matter covered herein. It is sold with the clear understanding that the Publisher is not engaged in rendering legal or any other professional services. If legal or any other expert assistance is required, the services of a competent person should be sought. FROM A DECLARATION OF PARTICIPANTS JOINTLY ADOPTED BY A COMMITTEE OF THE AMERICAN BAR ASSOCIATION AND A COMMITTEE OF PUBLISHERS.

Library of Congress Cataloging-in-Publication Data

Names: Pau, Tan Min, editor. | Ghaffar, Mazlan Abd, editor. | Abdullah, Muhd Danish-Daniel, editor. | Azra, Mohamad Nor, editor.
Title: Climate change and environmental perturbations : impacts on biodiversity / Tan Min Pau, Mazlan Abd. Ghaffar, Muhd Danish-Daniel Abdullah, Mohamad Nor Azra, editors.
Description: New York : Nova Science Publishers, [2024] | Series: Marine science and technology | Includes bibliographical references and index.
Identifiers: LCCN 2023051910 (print) | LCCN 2023051911 (ebook) | ISBN 9798891132719 (paperback) | ISBN 9798891133761 (adobe pdf)
Subjects: LCSH: Marine ecosystem health--Research--Malaysia. | Mariculture--Climatic factors--Malaysia. | Marine animals--Climatic factors--Malaysia. | Marine animals--Climatic factors--Malaysia. | Climatic changes--Research--Malaysia.
Classification: LCC QH185 .C55 2024 (print) | LCC QH185 (ebook) | DDC 338.3/71809595--dc23/eng/20240102
LC record available at https://lccn.loc.gov/2023051910
LC ebook record available at https://lccn.loc.gov/2023051911

Published by Nova Science Publishers, Inc. † New York

Contents

Preface

Climate change and human-induced environmental perturbations have become rather common phenomena in recent decades, and it is crucial that they are tackled and mitigated as soon as possible. In aquatic environments, climate change affects aquatic organisms through changes in water temperature, pH, nutrients, salinity, and sea level, among other factors, subsequently impacting the adaptation, survival, and sustainability of species. In addition, anthropogenic activities that cause environmental problems, such as mercury pollution and pesticide run-off, further threaten the well-being of aquatic organisms. Recognizing the importance of novel knowledge and research innovations in pursuing our nation's agenda regarding sustainable development goals, we have taken the initiative to present this book. Through this book, we are delighted to detail some of the research activities conducted by the researchers at Institute of Climate Adaptation and Marine Biotechnology (ICAMB), Universiti Malaysia Terengganu (UMT). The chapters presented in this book are based on their fields of expertise, and comprise significant outputs from the research grants or research program at ICAMB, UMT.

The Editorial Team
Min Pau Tan, Mazlan Abd. Ghaffar, Muhd Danish-Daniel Abdullah and Mohamad Nor Azra

Introduction

Min Pau Tan and Siti Azizah Mohd Nor

"Global warming" and "climate change" have become common and trending subjects of interest in recent decades, due to their massive influence on biodiversity and subsequently affecting the sustainable usage of ecological resources by human beings. In recent times, various ecosystems have been severely inundated with issues which threaten the very survival of the biodiversity that we depend on. Biodiversity is essential, as our health, food, and economy depend on it. Unfortunately, the rapid change of the earth's climate and anthropogenic stressors are affecting all forms of life and non-life on earth, of which most such effects are irreversible. In this book, "*Climate change and environmental perturbation: Impacts on biodiversity*", a total of six chapters are presented. These chapters highlight the impacts of climate change and human-mediated environmental disturbances on biodiversity, as well as the use of micro-organisms in combating environmental pollution (including their potential as anti-biofouling agents).

In particular, in Chapter 1, this book begins with a systematic review and meta-analysis on climate change research in Malaysia, highlighting the challenges faced by our biodiversity as a result of climate changes and suggesting approaches that may address some of these issues. The impacts of climate change, such as increasing temperatures and flooding, affect every aspect of human life, from food production to the economy and health. This chapter describes the collation and meta-analysis of available research data in Malaysia, thus providing a comprehensive review of how climate change has impacted our local aquatic ecosystems, as illustrated in fish and shellfish species. Climate change-related events and patterns are shown to significantly affect the physiology, behaviour, distribution, performance, and health of aquatic animals. Such information is important for developing effective habitat, ecosystem, and resource management strategies, and may help relevant agencies to plan and implement coastal and inland fisheries management policies through the identification of climate change-vulnerable fish or shellfish species in different regions of Malaysia. Among others, the authors recommend that policymakers should focus on protecting "ecological engineers," such as corals, in order to improve the resilience of other species dependent on them, thus improving the health of aquatic animals in Malaysia.

While we acknowledge that the effects of climate change will not disappear, all is not lost if we are cognisant of this fact and implement immediate and strategic steps based on sound systematic research to contain the associated problems.

Chapter 2 extends our knowledge on the influence of historical climatic and geological arrangements on the present-day genetic distribution among fish species through a comprehensive review of the existing literature. This very interesting theory has been elucidated through extensive use of molecular and bioinformatic tools in a plethora of studies. The latter are computer software programs that have been developed to extract meaningful information from biological databases, facilitating better understanding of the events in the Earth's history—particularly during the Pleistocene and Pliocene Epochs—that have influenced the current genetic pattern of many species. The review highlights that even co-inhabiting species may respond differently. Thus, understanding the adaptive response of each organism to these extrinsic and intrinsic factors, and how they influence population segregation, is crucial to generate expectations regarding their responses to future sea level increases and, subsequently, promote the planning of management policies. However, any management strategy is contingent on accurate species identification. Failure to utilise the correct species or taxon group due to morphological ambiguity may not produce the desired effect, be it in an aquaculture, restocking, or fisheries programme, and could even be detrimental. Morphological traits are often influenced by both external (e.g., temperature, pH, salinity) and internal (e.g., life stage) environmental factors.

Chapter 3 highlights several dinoflagellates that are notoriously linked with harmful algae blooms (HABs) and many bacterial species that interact with this planktonic group, regulating the toxicity levels of the HABs.

Chapter 4 discusses the impacts of widespread mercury pollution caused by anthropogenic activities, such as fuel combustion and mining on land, which consequently affects aquatic habitats. This substance is toxic and its bioaccumulation and persistence are detrimental to the environment. Contaminated water is ingested by the various organisms, leaving consumers of the contaminated seafood at risk of mercury poisoning. This chapter describes the analytical methods which are utilised to evaluate the status of mercury contamination in the Malaysian seafood industry, as well as in different aquatic habitats.

Chapter 5 illustrates the utilisation of microbes as natural inhibitory materials to negate marine environmental pollution. It is well-known that millions of tonnes of pesticides are released into the environment each year to support agricultural activities, leading to high environmental pollution. Acknowledging this challenge, Chapter 5 describes the utility and efficacy of environmentally friendly bioremediation approaches developed through the use of micro-organisms such as Pseudomonas spp., Streptomyces spp., or Serratia spp. to degrade these insecticide pollutants. This comprehensive review, based on the literature spanning a period of 20 years, explains the sources, taxa, and approaches used to optimise the bioremediation process which could be further improved through more advanced technologies.

Chapter 6 takes us further into the use of micro-organisms to combat industrial biofoulers. These are unpleasant complex layers that form on the surfaces of materials in contact with seawater from bacteria, fungi, phytoplankton, barnacles, limpets, oysters, algae, and so on. If not efficiently addressed, biofouling can be highly detrimental to marine-based sectors such as shipping, power plants, aquaculture, and mariculture. This review, based on publications from 2011 to 2021, describes anti-biofouling bioactive compounds in numerous micro-organisms which have the potential to inhibit the growth of biofoulers, thus saving the relevant industries from high monetary losses.

In a nutshell, the negative impacts of climate change and anthropogenic activities on our local marine aquatic organisms are already being felt, as illustrated here. Moving forward, we must work together through specific and structured programmes to manage both ecosystems and communities to adapt and become more resilient in the face of these challenges. We close out this introduction by citing a relevant famous and impactful quote:

"*We do not inherit the Earth from our ancestors, we borrow it from our children*"

(Chief Seattle)

Chapter 1

Research Agendas on Marine Aquatic Animals and Climate Change Research in Malaysia: A Systematic Review and Meta-Analysis

Mohamad N. Azra[1,2,*]
Mohd Iqbal Mohd Noor[3,4]
Amiruddin Ahmad[1,5]
Tan Chun Hong[5,6]
Nadiah W. Rasdi[1,7]
Yeong Yik Sung[2]
Zainudin Bachok[6]
Liew Hon Jung[8]
Wee Hin Boo[9]
and Mazlan Abd Ghaffar[1,5,8]

[1]Institute of Climate Adaptation and Marine Biotechnology (ICAMB), Universiti Malaysia Terengganu, Kuala Nerus, Terengganu, Malaysia
[2]Research Center for Marine and Land Bioindustry (Earth Sciences and Maritime), National Research and Innovation Agency (BRIN), Pemenang, West Nusa Tenggara, Indonesia
[3]Faculty of Business Management, Universiti Teknologi MARA (UiTM) (Pahang), Raub, Pahang, Malaysia
[4]Institute for Biodiversity and Sustainable Development, Universiti Teknologi MARA (UiTM), Shah Alam, Selangor, Malaysia
[5]Faculty of Science and Marine Environment, Universiti Malaysia Terengganu, Terengganu, Malaysia

[*] Corresponding Author's Email: azramn@umt.edu.my.

In: Climate Change and Environmental Perturbations
Editors: T. Min Pau, M. Abd. Ghaffar, M. Danish-Daniel Abdullah et al.
ISBN: 979-8-89113-271-9

[6]Institute of Oceanography and Environment (INOS),
Universiti Malaysia Terengganu, Kuala Nerus, Terengganu, Malaysia
[7]Faculty of Fisheries and Food Sciences, Universiti Malaysia Terengganu,
Kuala Nerus, Kuala Terengganu, Malaysia
[8]Higher Institution Centre of Excellence (HICoE),
Institute of Tropical Aquaculture and Fisheries (AKUATROP),
Universiti Malaysia Terengganu, Terengganu, Malaysia
[9]Institute of Climate Change, Universiti Kebangsaan Malaysia (UKM),
Bangi, Selangor, Malaysia

Abstract

Systematized research is critical for long-term biodiversity management of aquatic animals and climate change. Malaysia is surrounded by diverse freshwater and marine ecosystems that support various functions such as artisanal fisheries and aquaculture activities and nursery ground for fish and shellfish species, which further support offshore fish and shellfish populations and fisheries. Freshwater ecosystems such as rivers, peat swamps, reservoirs and marine ecosystems such as estuaries, lagoons, mangroves which are more vulnerable to be impacted by climate change related issues. Hence, we aimed to develop a systematic review of the impact of climate changes on the aquatic fish and shellfish species, by characterizing the chemical, physiological and behavioural changes. Our analysis found that the proposed research field of aquatic animals and climate change requires more in-depth scientific study; only 22 articles were related with the proposed research area, with substantial representation across the themes of aquatic environment and climate change. There were significant three-way relationships between the factors of fish and shellfish species, climate change issues, and category of data ($P<0.05$). Fish-only research was primarily concerned with environmental factors, and there was a clear association between aquatic animals and data categories, as well as environmental factors studies focusing on behavioural data. This chapter identifies one of the emerging prospective future study areas. Development and research on aquatic ecosystems and fisheries for fish and shellfish species that are less sensitive to climate change, as well as assisting aquatic ecosystems and communities in preparing for change and becoming more resilient to the effects of climate change, are strongly encouraged.

Keywords: anthropogenic stressor, behavioural changes, coastal zones, distribution, environmental factors, global warming, physiological changes, pollution

Introduction

General Introduction

Understanding and identifying fish and/or shellfish species which are vulnerable to climate change is crucial in strengthening fisheries management and relieving the strain on fisheries caused by climate change. Malaysia has approximately 45,459 km^2 of freshwater ecosystems, including rivers, reservoirs, and mining pools. In addition, the country is surrounded by 42,330 km^2 of coastal marine ecosystems, such as estuaries, lagoons, mangroves, seagrass beds, coral reefs, and mudflats. These ecosystems play a significant role in carbon sequestration, which has been estimated as one billion metric tons of carbon per year (Yusoff et al., 2006; Silvius and Schrier, 2016; Charles et al., 2020; Gilbert et al., 2021).

The Food and Agriculture Organization (FAO) of the United Nations (UN) defines aquatic[1] as "Living in or on water for all or a substantial part of the life span," which mostly refers to animals or plants. A habitat is defined by the FAO as "Part of an ecosystem with conditions in which an organism naturally occurs or can establish," and it is popularly known as the natural place of occurrence of an organism. Aquatic organisms such as fish and shellfish are among the most important aquatic livestock but, unfortunately, they are threatened by climate change.

Aquatic habitats contain large numbers of fish and shellfish species, from the world's smallest fish (the dwarf minnow of the genus *Paedocypris*) and world's tiniest shellfish crustacean (*Stygotantulus stocki*), to the largest fish (the whale shark, *Rhincodon typus*) and shellfish (giant clams of the genus *Tridacna*) in the world. In this study, we refer to fish in terms of finfish, which FAO defines as a cold-blooded lower vertebrate that (usually) has fins, gills, and scales, and which lives in water. Shellfish are referred to as aquatic invertebrates possessing a shell or exoskeleton, comprising molluscs (e.g., oyster, clams, cockles, and mussels) and crustaceans (e.g., lobsters, prawn, shrimp, and crabs). Fish and shellfish occur in fresh, brackish, and marine waters. The FAO also defines other important terms related to current studies of fresh, brackish, and marine waters. Freshwater is termed as "Water occurring on the earth's surface in glaciers, lakes and rivers (i.e., surface water), and underground in aquifers (i.e., groundwater). Its key factor is a low concentration of dissolved salts." Brackish water is defined as "Water in which

[1] Term available at http://www.fao.org/faoterm/en/.

the salinity is appreciable but not to a constant high level, usually showing wide salinity fluctuations. It is often characterized by regular daily and seasonal fluctuations in salinity due to freshwater and full-strength marine water influxes." Marine water is defined as "Coastal and offshore waters in which the salinity is maximal (around 35 ppt) and not subjected to significant daily and seasonal variation." Freshwater (river), brackish water, and marine aquatic habitats are distinct environments supporting a variety of natural functions, such as flood mitigation and irrigation for agricultural purposes, artisanal fisheries and aquaculture activities, and nursery grounds for fish and shellfish species, including those for offshore fish and shellfish populations, thus supporting related fisheries (Nagelkerken et al., 2008; Rosli and Zain, 2016).

Recent studies have revealed that these water environments are vulnerable to climate change, particularly in terms of extreme seasonal variations, El Nino and La Nina events, and salinization/marine water intrusion due to sea level rise (Tangang et al., 2012; Matias et al., 2021). Previous studies have also shown that climate change-related events and patterns could significantly affect the physiology, behaviour, distribution, performance, and health of aquatic organisms, especially primary and secondary producers such as plankton, fish, and shellfish within these ecosystems (Adnan et al., 2002; Ooi and Chong, 2011; Yusof et al., 2012; Yurimoto et al., 2014; Abdullah et al., 2015a; Abdullah et al., 2015b; Mirzaei and Hwai, 2015; Moh et al., 2015; Mokhtari et al., 2015; Ara et al., 2016; Ikhwanuddin et al., 2016; Zieritz et al., 2016; Mirzaei et al., 2017; Sule et al., 2018; Wilkinson et al., 2018; Yusof et al., 2018; Wilkinson et al., 2019; Yeny-Nadira et al., 2019; Zakariah et al., 2019; Ara et al., 2020; Kadir et al., 2020; Mukhtar et al., 2020; Yuslan et al., 2021).

Climate change affects the whole global ecosystem, with Southeast Asia expected to be one of the hardest-hit regions by the end of the 21st century. This is due to the velocity of climate change in the equatorial regions (including Southeast Asia) and the shift in biodiversity in this biologically diverse area (Burrow et al., 2011). Temperatures in Southeast Asian countries such as Malaysia are expected to fluctuate and rise erratically (Griffiths et al., 2005; Tan et al., 2021), accompanied by extreme rainfall events (Walsh et al., 2013; Tan et al., 2017), haze pollution disasters (Aghamohammadi and Isahak, 2017), sea level rises (Ercan et al., 2012), and extreme weather events such as flooding, droughts, and heat waves (Tangang et al., 2012; Tang, 2019). Climate change review studies on agriculture and food security issues have been conducted in Malaysia (Tang, 2019; Firdaus et al., 2020). However, there has been no investigation relating climate change to aquatic animals, particularly fish and shellfish species, based on an extensive literature search. This limits our understanding of the effects of climate change on commercially and economically important aquatic animals such as fish and shellfish. Thus, research on the impact of climate change is critical for understanding the threats related to aquatic agricultural research and for better strategizing policymaking.

This chapter collates available research data obtained in Malaysia, in order to provide a comprehensive review of how climate change has impacted aquatic fish and shellfish species. Assessing the impact of climate change on aquatic animals, particularly fish and shellfish species, is important for effective habitat, ecosystem, and resource management and, therefore, is highlighted here. Furthermore, unique information that is critical to the survival and development of these species under future climate change scenarios is discussed. The impact of climate change on aquatic fish and shellfish species is evaluated by analysing chemical, physiological, and behavioural changes. This review concludes with the identification and presentation of emerging potential future research areas.

Stakeholder Involvement

This scope and objective of this systematic review are based on a research programme funded by the Malaysian government. Four local higher learning education institutes—(i) Universiti Malaysia Terengganu (UMT), (ii) Universiti Malaya (UM), (iii) Universiti Sains Malaysia (USM), and (iv) International Islamic University Malaysia (IIUM)—were involved in the research programme entitled "Ocean Climate Change Potential Risk, Impact and Adaptation Towards Marine and Coastal Ecosystem Services in Malaysia," with UMT as the leading institution of the four-year sponsored programme. A systematic literature review was conducted to identify previous and current climate change impacts on the aquatic ecosystem, especially in Malaysia. This review can help policymakers to plan strategies for coastal and inland fisheries management, through the identification of climate change-vulnerable fish or shellfish species in different regions of Malaysia for future policy recommendations. All authors were involved in the initial online meetings and discussions conducted to develop the aims and scope of the study.

Objective of The Review

We aimed to create a systematic literature review focused on the impacts of climate change on aquatic fish and shellfish species, based on their various chemical, physiological, and behavioural changes. The primary research question for this review is: "What is the impact of climate change on Malaysian aquatic ecosystems?" This research question was structured through a PEO (population, exposure, outcome) approach, as detailed in Table 1.

Table 1. Components of primary research questions

Population	Exposure	Outcome
Malaysia Aquatic Ecosystem	Climate change	Fish and shellfish chemical, physiological, and behavioural changes

Methods

Search Terms and Strings

The guidelines in "Collaboration for Environmental Evidence and the Reporting Standards for Systematic Evidence Syntheses" (ROSES) were used in this study (Haddaway et al., 2017). The ROSES form is included as an additional file. The search strings used in this review included a combination of the keywords "Malaysia," "Marine" and "Climate Change" (Table 2). The Web of Science (WOS) core collection and Scopus database were used for the search, which are two of the most-used databases for systematic reviews (Azzeri et. al. 2020). The search terms for the two main keywords were combined by using the Boolean operator "OR" and "*" for single characters. Recorded references were imported into an EndNote library, where all duplicates were removed and their numbers were recorded.

Table 2. List of terms, keywords, and search strings used in the present study

Terms used	Keyword	Search string
Population terms	Malaysia	(Malaysia)
	Marine	(marine) OR (ocean) OR (sea) OR (coast*) OR ("coastal zone") OR ("coastal water*") OR (shore) OR ("offshore water*") OR (intertidal) OR (aquatic) OR (salinit*) OR (seawater)
Exposure terms	Climate change	(("climat* chang*") OR ("climat* risk*") OR ("climat* effect*") OR ("climat* variab*") OR ("climat* variat*") OR ("climat* extrem*") OR ("climat* uncertaint*") OR ("climat* hazard*") OR ("climat* disaster*") OR ("climat* shock*") OR ("climat* stressor*") OR ("climat* impact*") OR ("climat* condition*") OR ("global warm*") OR ("global climat*") OR ("global chang*") OR ("seasonal* variat*") OR ("weather* variab*") OR ("extreme* event*") OR ("environment* variab*") OR ("environment* impact*") OR ("environment* chang*") OR ("environment* factor*") OR ("anthropogenic effect*") OR ("temperature ris*") OR ("temperature effect*") OR ("sea level chang*") OR ("sea level ris*") OR ("el-nino") OR ("la$nina") OR (drought*) OR (precipit*) OR (rainfall*) OR ("carbon dioxide") OR ("greenhouse gas") OR ("greenhouse effect*") OR ("air temperat*") OR ("atmospheric temperat*") OR ("sea surface* temperat*") OR (runoff) OR ("melt* of the glacier*"))

Article Screening and Eligibility Criteria

Article Screening

Based on the ROSES method, the chapter were evaluated using two approaches: (i) Title and Abstract, and (ii) Full text. Two of the reviewers (MA and MI) evaluated the articles independently in the first category, while three reviewers evaluated those in the second category (MA, MI, and WHB). Then, information from the articles that were deemed eligible were extracted, based on their design, impact measures, and reported outcomes, in a standardized evidence table. To ensure consistency, the evaluation data were compared, and any conflicts in opinion between reviewers were resolved through discussion before the screening could proceed. The original screening and validity appraisal criteria were reviewed to reconcile disagreements during Kappa analysis, at each level of screening and among all the reviewers (Belur et al., 2021; Park and Kim, 2015).

Study Inclusion Criteria

The final criteria to include the articles were as follows:

Inclusion criteria: (i) Aquatic animals, focused on fish and shellfish, (ii) Direct climate change impacts, (iii) Study carried out in Malaysia, and (iv) Research article with data.

Exclusion criteria: (i) Aquatic plants, single-cell organisms, phytoplankton, zooplankton, or coral; (ii) non-climate change-related issue (clearing activities for development projects, construction waste runoff, bauxite residues runoff or flood mitigation works); (iii) Articles which include experimental data from other countries; and (iv) Life cycle assessment article, prediction or forecasting-based assessment(s), laboratory experiments (indirect impacts), or review article.

Validity Assessment of Study

Articles included in the full text category were then evaluated using standard quality assessment criteria for evaluating quantitative research adapted from Azzeri et al., (2020). The checklist included eight (8) assessment criteria; namely, (1) study objectives, (2) study designs, (3) study outcomes, (4) sample size calculations, (5) analysis of findings, (6) variance estimates for the main results, (7) reporting sufficient results, and (8) conclusions. The articles were given two points if they satisfied the underlying criteria, one point if they partially satisfied the criteria, and zero if they did not. Following this procedure, the maximum quality score for a paper is 16. To be included, articles had to attain a quality standard of more than ten (10).

Data Extraction and Synthesis

For this study, we adapted the framework of Goulas et al., (2020) and Häkkilä et al., (2021) for data extraction and synthesis. The extracted data were organized into two categories: (i) Descriptive/narrative, and (ii) meta-analysis. Studies that provided quantitative data on outcome measures were included in the meta-analysis. Studies that had incomplete or missing information that could not be retrieved were excluded from the meta-analysis.

Descriptive/Narrative

The descriptive analysis focuses on various aspects of the articles, such as title of the publication, article citation, year of publication, country of lead author, and institutional affiliation. Then, a narrative synthesis is carried out by using a thematic category that revolves around each article's key findings. These mainly include the study hypotheses; in this case, explaining the impact of climate changes on aquatic ecosystems. The synthesis followed a two-step coding process: first, a line-by-line review of the articles was conducted to identify the key impacts on aquatic ecosystems. A code label was then assigned to each article, in order to cluster all of the information into a common theme. Two of the reviewers (MA and MI) independently performed the coding process, and discussion was conducted among all of the reviewers to create a robust set of common themes.

Meta Analysis

Research papers on the impacts of climate change on aquatic animals in Malaysia were compiled and organized according to the following nine details: State, Region, Aquatic animals, Water bodies, Climate change issue, Sampling period, Category of data, Year published, and Authors (Table 3). Six factors (independent variables) were chosen from the details (n = number of categories): Region (n = 4), Aquatic animals (n = 3), Water bodies (n = 3), Climate change issue (n = 3), Sampling period (n = 3), and Category of data (n = 3). Meanwhile, Year published (2002–2020) was selected as the dependent variable.

Table 3. Details of data collected regarding the impact of climate change on aquatic animals

State	Region	Aquatic animals	Water bodies	Climate change issue	Sampling period	Category of data	Year published	Author(s)
Perak	West coast	Fish & Shellfish	Brackish water	Anthropogenic activities	< 1 year	Chemical changes	2015	Abdullah et al., (a)
Perak	West coast	Fish & Shellfish	Brackish water	Anthropogenic activities	< 1 year	Chemical changes	2015	Abdullah et al., (b)
Johor	West coast	Fish & Shellfish	Brackish water	Anthropogenic activities	< 1 year	Chemical changes	2020	Mukhtar et al.
Terengganu	East coast	Shellfish	Marine	Climatic changes	≥ 1 year < 2 years	Physiological changes	2019	Zakariah et al.
Penang	West coast	Shellfish	Marine	Environmental factors	< 1 year	Physiological changes	2015	Mirzaei and Hwai
Selangor	West coast	Shellfish	Marine	Climatic changes	< 1 year	Physiological changes	2014	Yurimoto et al.
Penang	West coast	Shellfish	Brackish water	Environmental factors	≥ 1 year < 2 years	Physiological changes	2017	Mirzaei et al.
Selangor/ Negeri Sembilan	West coast	Shellfish	Brackish water	Environmental factors	< 1 year	Behavioural changes	2015	Mokhtari et al.
Selangor	West coast	Shellfish	Brackish water	Environmental factors	≥ 1 year < 2 years	Behavioural changes	2015	Moh et al.
Perak	West coast	Fish	Freshwater	Environmental factors	< 1 year	Behavioural changes	2018	Yusof et al.
Perak	West coast	Fish	Freshwater	Anthropogenic activities	< 1 year	Behavioural changes	2016	Ikhwanuddin et al.

State	Region	Aquatic animals	Water bodies	Climate change issue	Sampling period	Category of data	Year published	Author(s)
All state (except Borneo)	West coast/ East coast	Shellfish	Freshwater	Environmental factors	≥ 1 year < 2 years	Behavioural changes	2016	Zieritz et al.
Johor	West coast	Fish	Brackish water	Environmental factors	≥ 1 year < 2 years	Behavioural changes	2016	Ara et al.
Sabah	Borneo	Fish	Freshwater	Anthropogenic activities	< 1 year	Behavioural changes	2018	Wilkinson et al.
Perak	West coast	Fish	Brackish water	Environmental factors	≥ 1 year < 2 years	Behavioural changes	2020	Ara et al.
Perak	West coast	Fish	Brackish water	Environmental factors	≥ 1 year < 2 years	Behavioural changes	2011	Ooi and Chong
Perak/ Selangor/ Negeri Sembilan/ Johor	West coast	Fish	Brackish water	Environmental factors	< 1 year	Behavioural changes	2012	Yusof et al.
Terengganu	East coast	Fish	Freshwater	Climatic changes	≥ 1 year < 2 years	Behavioural changes	2020	Kadir et al.
Sabah	Borneo	Fish	Freshwater	Climatic changes	≥ 2 years	Behavioural changes	2019	Wilkinson et al.
Terengganu/ Kelantan / Pahang	East coast	Fish	Marine	Environmental factors	≥ 2 years	Behavioural changes	2019	Yeny-Nadira et al.
Perak	West coast	Shellfish	Brackish water	Climatic changes	≥ 2 years	Behavioural changes	2002	Adnan et al.
Selangor	West coast	Fish	Freshwater	Environmental factors	< 1 year	Behavioural changes	2018	Sule et al.

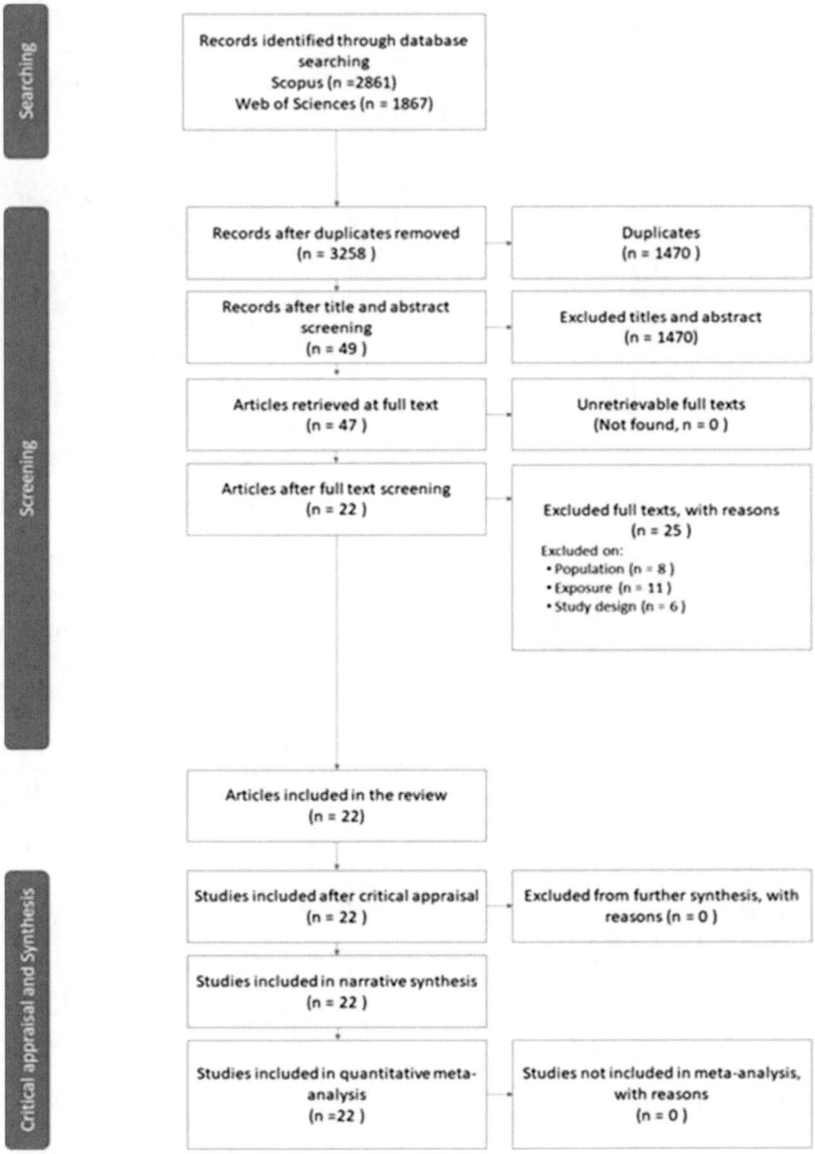

Figure 1. Flowchart of study selection based on the ROSES approach.

The data set was imported into R console V 4.1.0 using the RStudio V1.4.1106 user interface (Version of 2021: Rcore, RStudio). Normality tests (Shapiro–Wilk test, SW) and homoscedasticity tests (Fligner–Killeen test,

FK) were conducted on the independent factors for each Year published (dependent variable). To observe the patterns of impact on aquatic animals in the climate change studies conducted in Malaysia, median tests (Kruskal–Wallis test, KW) were carried out on each of the six independent variables, with respect to the year published. To test the relationships among the independent variables, Fisher's exact tests (FE) with 2000 permutations were conducted, due to the low number of data points ($n = 22$). The significance level of all tests was set at $\alpha = 0.050$.

Results

Study Flow

The flowchart depicting the study selection process is shown in Figure 1. Searches for relevant articles were conducted from 1–2 June of 2021, resulting in 4728 articles retrieved from the two databases (Scopus and WoS). Duplicates were removed, and the titles and abstracts of the articles were screened. After removing duplicates and screening titles and abstracts, 49 articles were found to be relevant. There were two articles that could not be found on the website. Subsequently, a full-text screening on the remaining articles was conducted. Articles that did not meet the inclusion and exclusion criteria were rejected. Finally, 22 articles were chosen for the systematic review after full-text screening. The most common reason for exclusion following full-text screening was failing to meet the full criteria (i.e., no direct climate change issue or data involving climate change, 11 articles). The list of articles included or excluded at the full-text stage is provided in an additional file, along with justification.

Review Findings

Narrative Synthesis

In total, 22 relevant publications relating to aquatic animals and climate change in Malaysia between 2002 to 2020 were obtained. The state and region with the highest number of such studies conducted were Perak ($n = 7$) and the West Coast of Peninsular Malaysia ($n = 16$), respectively. Sabah was the only state from Borneo included in this review, and no studies from Melaka or

Sarawak were found. Studies on fish (n =11, 50% of the total literature; Figure 2A) and brackish water (n = 10, 40.9% of the total literature; Figure 2B) were the most common. The majority of the climate change issues investigated were related to environmental factors (n = 12, 54.5% of the literature;

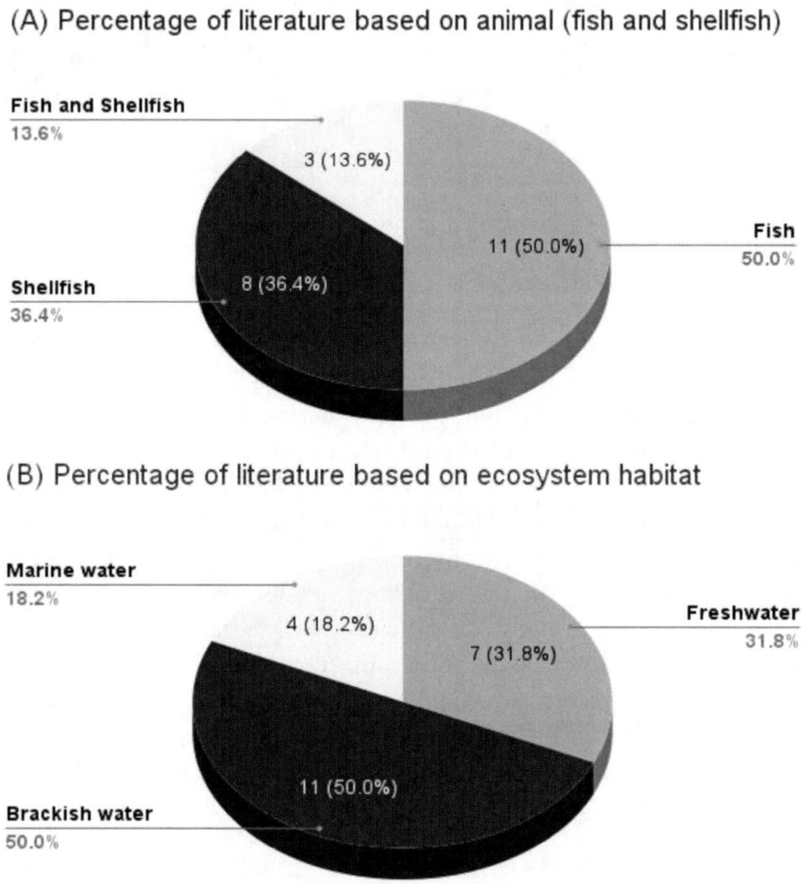

Figure 2. Percentage of literature based on: (A) Animal (fish, shellfish, or fish–shellfish mix); and (B) ecosystem habitat (freshwater, brackish water, or marine water).

Figure 3A), and most articles had a study period of less than 1 year (n = 11; 45.5% of the collected literature; Figure 3B). Behavioural studies (n = 15) accounted for 68.2% of the studies in the literature relating to aquatic animals and climate change (Figure 4). The Years published were not normally distributed (SW: $W = 0.800, p < 0.001$), and the variances were homoscedastic for all the factors (FK: $p > 0.050$). The median tests revealed no significant pattern in years published for the articles with each factor (KW: $p > 0.050$; Table 4). However, there were significant three-way relationships between the factors Aquatic animals, Climate change issues, and Category of data (FE: $p < 0.050$). Fish-only studies were focused mainly on Environmental factors ($n = 7$), while anthropogenic issues were conducted mostly on a fish and shellfish mixture ($n = 3$). On the other hand, there was a clear relationship between aquatic animals and category of data: Fish-only studies were focused on behavioural data ($n = 11$), fish and shellfish mixed studies were focused on chemical data ($n = 3$), and shellfish-only studies included both behavioural and physiological data ($n = 4$, respectively).

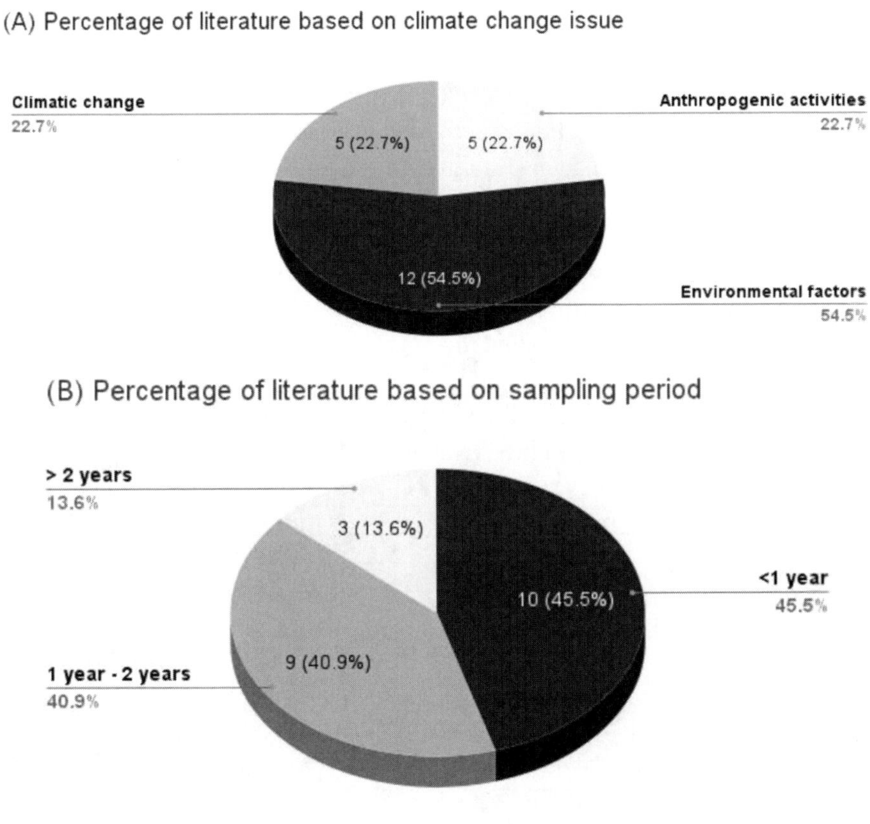

Figure 3. Percentage of literature based on: (A) Climate change-related issues; and (B) the sampling period.

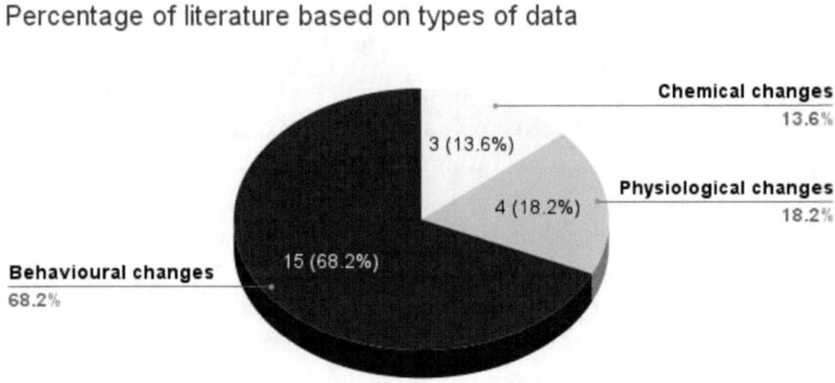

Figure 4. Percentage of data based on types of change due to climate changes.

Finally, there were clear environmental factor studies focused on behavioural data (n = 10). Furthermore, Region and Water bodies had a significant relationship (FE: p = 0.020), with brackish water studies mainly focused on West coast of Peninsular Malaysia (n = 10), marine organism studies generally in Peninsular Malaysia (n = 5), while Borneo (specifically, Sabah; n = 2) only focused on freshwater organisms (Table 5). On the other hand, there was a significant relationship between Water bodies and Category of data (FE: p = 0.017), with behavioural studies mainly focused on brackish and freshwater organisms (n = 7, respectively), while physiological studies were focused on marine organisms (n = 3).

Table 4. Meta-analysis regarding the impact of climate change on aquatic animals based on the factors considered in this study

Shapiro-Wilk test	W			p		
	0.800			0.001		
Type of test applied	Fligner–Killeen test			Kruskal–Wallis test		
Factors	X2	df	p	X2	df	p
Region	3.576	3	0.312	6.851	3	0.077
Aquatic animals	0.312	2	0.856	2.961	2	0.228
Water bodies	2.685	2	0.261	3.480	2	0.176
Climate change Issue	0.494	2	0.782	0.287	2	0.867
Sampling period	0.769	2	0.681	0.949	2	0.623
Category of data	0.741	2	0.691	0.119	2	0.943

Table 5. Relationships between factors considered in the study and the impacts of climate change on aquatic animals

	Region	Aquatic animals	Water bodies	Climate change issue(s)	Sampling period
Aquatic animals	0.849	-	-	-	-
Water bodies	**<u>0.020</u>**	0.355	-	-	-
Climate change issue	0.191	**<u>0.019</u>**	0.416	-	-
Sampling period	0.067	0.611	0.617	0.059	-
Category of data	1.000	**<u>0.000</u>**	**<u>0.017</u>**	**<u>0.022</u>**	0.502

Bold and underlined values denote significant differences.

Study Discussion

Table 6 provides a summary of the direct effects of climate change on aquatic fish and shellfish in Malaysia, and relevant significant findings are discussed. Table 7 summarizes the impacts of the data/parameters related to climate change on aquatic fish and shellfish in Malaysia, as well as the main data found (i.e., chemical, physiological, and behavioural) and the conclusions of the literature review.

Chemical Changes

Significant chemical changes have been found to harm aquatic fish and shellfish that live in areas where water pollution occurred as a result of anthropogenic activities including agricultural runoff (Abdullah et al., 2015a; Abdullah et al., 2015b; Mukhtar et al., 2020). Biochemical compounds, mercury pollution and contamination, radionuclides, and biocides are consumed by fish and shellfish that live near the seafloor, where toxic substances accumulate as evidenced by their deposition in the sediment. Furthermore, these studies cautioned that such compounds could have a downstream effect on human health, as measured through the health risk index. More research is required to examine the effects of chemical alterations of substances on resident or cultured fish and/or shellfish, or those close to coastal areas where agricultural activities are more intensive. Recent statistics issued by the Department of Statistics of Malaysia found that oil palm was the main contributor—at 37.7% of the agriculture sector—with most planted near the coastal areas of Malaysia (DoS, 2020). The previous study by Gopalakrishnan et al., (2019) emphasized that agricultural activities near the coast readily lead to climate change-related problems such as agricultural runoff, where wastes from agriculture, through the excessive use of fertilizers and pesticides, release toxic pollutants into the water, reducing fish production nearby. A study conducted by Ali et al., (2018) has revealed that fish larvae of fathead minnows (*Pimephales promelas*) were exposed to contaminated water and sediment from agricultural runoff events at Elkhorn River, Nebraska, USA. They showed that the agriculture runoff significantly caused contamination of the fish living near agricultural activity areas (Table 6). Agricultural runoff also led to higher concentrations of glyphosate, nitrate, and ammonia at a cockle farming area in Bagan Pasir, Perak, Malaysia (Joni et al., 2021). Chemical compounds such as methyl mercury, thorium, uranium, radium-226, radium-228, potassium-40, organotin, and biocides are introduced into the soil through agricultural activities, then discharged into the marine environment with runoff (Abdullah et al., 2015a; Abdullah et al., 2015b; Mukhtar et al., 2020). With the increased frequency of climate change-driven events, it is expected that the biochemical compositions of the animals and their habitats will be continuously impacted, potentially impacting the health of the sampled animals as well as human health, through consumption of these organisms.

Physiological Changes

Physiological changes, such as growth, mass mortality, and significant intensity of protozoan parasites, have been associated with climate change-related environmental factors, such as heavy rainfall, temperature, and salinity (Yurimoto et al., 2014; Mirzaei and Hwai, 2015; Mirzaei et al., 2017; Zakariah et al., 2019); see Table 6. The shell growth of marine blood cockles (*Anadara granosa*) was significantly affected by the temperature and salinity fluctuation of the sea (Table 7). These abiotic stressors may act synergistically to affect physiological changes in the species. Higher metabolism was triggered by higher sea temperatures, which thus increased the maximum width of the cockle (Mirzaei and Hwai, 2015). Temporal variation is one of the contributing factors affecting the shell growth of *A. granosa* at different geographic locations due to different sea temperatures (Mirzaei et al., 2014; Ong et al., 2017). Previous research has also revealed that water temperature is one of the main abiotic factors influencing the physiological growth response of other aquatic invertebrates such as crustaceans (Azra et al., 2020). Mirzaei and Hwai (2015) discovered that temperature—rather than salinity—significantly influenced cockle shell growth. Nonetheless, salinity has been found to be a factor influencing the spawning season of *A. granosa* (Mirzaei et al., 2017). Furthermore, previous research has revealed that salinity is a problem for most bivalve species, reducing activity and energy acquisition while increasing energy demand to avoid osmotic shock (Peteiro et al., 2018).

Table 6. Direct impacts of climate change on aquatic fish and shellfish in Malaysia

No.	Fish or/and Shellfish (Common name, Scientific name)*	Climate change issue	Area and location	Significant finding(s) and Discussion	Sampling period	References
1.	Fish (Bearded Croaker, *Johnius dussumieri*; Flatfish, *Pseudorhombus malayanus*; Spotted Catfish, *Arius maculatus*) & Shellfish (Blue swimming crab, *Portunus pelagicus*; Ridged swimming crab, *Charybdis natator*)	Agricultural runoff	Coastal water area & Manjung, Perak	Highest mercury concentration was in *P. pelagicus* for both seasons, caused by the fact that the blue swimming crabs live and feed at the bottom of sea, where various kinds of toxic elements accumulate and deposit in the sediments.	Dry (March to June 2013) and rainy (July to September 2013) seasons (< 1 year)	Abdullah et al., 2015a
2.	Fish (Bearded Croaker, *Johnius dussumieri*; Flatfish, *Pseudorhombus malayanus*; Spotted Catfish, *Arius maculatus*) & Shellfish (Blue swimming crab, *Portunus pelagicus*; Ridged swimming crab, *Charybdis natator*)	Anthropogenic activities	Mouth river to coastal water & Bekah River, Manjung, Perak	a) Flatfish indicated the highest activity concentration of radionuclides. *P. malayanus* feeding and living on or near the bottom of sediment, gravel/rocks, muddy, or sandy areas, where hazardous and toxic substance are deposited and accumulated. b) Radionuclides are higher in dry than rainy season, related to metabolism increment due to evaporation caused by increased temperature in dry season.	Dry (March to June 2013) and rainy (July to September 2013) seasons (< 1 year)	Abdullah et al., 2015b

Table 6. (Continued)

No.	Fish or/and Shellfish (Common name, Scientific name)*	Climate change issue	Area and location	Significant finding(s) and Discussion	Sampling period	References
3.	Fish (Jewfish croaker, *Otolithus ruber*; Blackspot threadfin, *Polydactylus sextarius*; Dussumier's anchovy, *Thryssa dussumieri*) & Shellfish (Pearl conch snail, *Strombus turturella*; Asian green mussels, *Perna viridis*; Fan mussel, *Pinna nobilis*)	Agricultural runoff	Estuary area, Gelang Patah, Johor	Asian green mussels and Fan mussels contained the highest level of biocides; namely, phenyltins (PTs). PTs were directly ingested from the polluted sediment through agricultural discharge by water runoff (Agricultural runoff).	Within 2018–2019** (< 1 year)	Mukhtar et al., 2020
4.	Shellfish (Ark shell, *Anadara cornea*)	Temperature, salinity, heavy rainfall	Lagoon area, Setiu, Terengganu	Prevalence and intensity of protozoan parasites (Gregarines) slightly higher during the rainy season. This was significantly related to the monsoon season, affected by the heavy rainfall regime.	January to December 2018 (1 year)	Zakariah et al., 2019
5.	Shellfish (Blood Cockle, *Anadara granosa*)	Sea temperature, salinity, rainfall	Island area & Balik Pulau, Penang	i) Reliable relationships between shell growth and environmental factors of sea temperature and salinity.	Mid-December 2011 and the end of January 2012 (< 1 year)	Mirzaei and Hwai, 2015

No.	Fish or/and Shellfish (Common name, Scientific name)[*]	Climate change issue	Area and location	Significant finding(s) and Discussion	Sampling period	References
6.	Shellfish (Blood cockles, *Anadara granosa*)	Heavy rainfall	Coastal area of Selangor	Mass mortality (>30%) occurrence caused by intensive precipitation,	January and March 2012 (< 1 year)	Yurimoto et al., 2014
7.	Shellfish (Blood cockles, *Anadara granosa*)	Temperature, salinity	Intertidal site (i.e., coastal area), Balik Pulau, Penang	Number of microgrowth lines was significantly different from intertidal areas. Different environmental conditions showed shell growth decreased when salinity decreased, caused by salinity fluctuation.	December 2011 to November 2012 (1 year)	Mirzaei et al., 2017
8.	Shellfish (Fiddler crab: *Uca paradussumieri, U. rosea, U. triangularis, U. forcipata, U. perplexa, U. annulipes and U. vocans*)	Temperature, salinity, pH	Mangrove forests, Sepang/Lukut, Selangor & Negeri Sembilan	Crab densities (among study sites) significantly differed among the various environmental parameters. Environmental factors related to climate change impacted habitat and sediment, which consequently affected crab density.	Within 2013–2014[**] (< 1 year)	Mokhtari et al., 2015
9.	Shellfish (Mud lobsters, *Thalassina anomala, T. kelanang* and *T. gracilis*)	Salinity	Estuarine, mangrove area and small coastal inlet, Port Klang, Selangor	The distribution of *T. kelanang* was correlated with salinity of seawater. High salinity caused *T. kelanang* to colonize the relatively sandy substrate near the seawater area.	October 2010 to March 2012 (> 1 year)	Moh et al., 2015

Table 6. (Continued)

No.	Fish or/and Shellfish (Common name, Scientific name)*	Climate change issue	Area and location	Significant finding(s) and Discussion	Sampling period	References
10.	Fish (10 species)***	Agricultural runoff, heavy rainfall, high temperature	Stream area, Bagan Datuk, Perak	A strong relationship between fish abundance and dissolved oxygen concentration was recorded. This explained the high number of fish recorded around the selected sampling site, which had high dissolved oxygen.	December 2017 (< 1 year)	Yusof et al., 2018
11.	Fish (21 species)***	Seasonal variation, anthropogenic activities	Upper river area, Kuala Kangsar, Perak	In the upper river area, the mean of COD was quite low due to anthropogenic activities.	Dry (June 2014), wet (August 2014) and moderately wet seasons (April 2015) (< 1 year)	Ikhwanuddin et al., 2016
12.	Shellfish (Freshwater mussels, 10 different species***)	Acidification, eutrophication	Freshwater habitat, Peninsular Malaysia***	Number of freshwater mussel species differed significantly from their water quality requirements. The mussel distribution was strongly affected by the geological conditions characterised by granite bedrock in inland & upland areas.	November 2014 and October 2015 (1 year)	Zieritz et al., 2016

No.	Fish or/and Shellfish (Common name, Scientific name)*	Climate change issue	Area and location	Significant finding(s) and Discussion	Sampling period	References
13.	Fish (Silver moony, *Monodactylus argenteus*; Glassfish, *Ambassidae* sp.; Ponyfish, *Leiognathidae* sp.; Banded trumpeter, *Terapon theraps*; Damselfish, *Pomacentridae* sp.)	Temperature, salinity, pH	Mangrove, estuary & coastal area, Gelang Patah, Johor	Two species, *Ambassidae* sp. And *Leiognathidae* sp., correlated significantly with salinity. Significant spatio-temporal variations were observed for salinity, which might influence the assemblage of larval fish.	October 2007 to September 2008 (1 year)	Ara et al., 2016
14.	Fish (34 species***)	Extensive anthropogenic activities	Headwater streams, Southeastern Sabah	Extensive agriculture-related activities significantly reduced local fish species richness. Stream sites within plantations lacking riparian reserves presented decreased fish species richness.	February to July 2017 (< 1 year)	Wilkinson et al., 2018
15.	Fish (14 different families, as the larval stage of certain fish did not allow for identification of its species name)	Temperature, salinity, pH	Mangrove area, Kuala Sepetang, Perak	Positive significant correlation between the fish family of Blenniidae and water temperature. Low temperature increases dissolved oxygen level, consequently increasing fish abundance at the sampling station.	April 2015 and March 2016 (1 year)	Ara et al., 2020

Table 6. (Continued)

No.	Fish or/and Shellfish (Common name, Scientific name)[*]	Climate change issue	Area and location	Significant finding(s) and Discussion	Sampling period	References
16.	Fish (19 families, as the larval stage of certain fish did not allow for identification of its species name)	Salinity	Mangrove and coastal area, Kuala Sepetang, Perak	The abundance of ambassid larvae in offshore areas, for all ontogenetic stages, were significantly ($P < 0.05$) higher than inside the mangrove estuary. This could have been caused by the consistent monthly mean salinity in offshore waters; whereas, in the mangrove, salinity ranged from 15.4 to 27.8 ppt.	May 2002 and October 2003 (> 1 year)	Ooi and Chong, 2011
17.	Fish (Javanese medaka, *Oryzias javanicus*; Indian medaka, *Oryzias dancena*)	Salinity	Estuary area, Peninsular Malaysia[***]	*O. javanicus* inhabited environments with higher salinity. Coexistence was found only at two of 12 sampling points. Co-existence of both species was the result of brackish water areas under the influence of tidal movements.	January until December 2009 (< 1 year)	Yusof et al., 2012
18.	Fish (31 species[***])	Temperature, salinity and rainfall	River mouth, Setiu, Terengganu	No significant difference was observed between fish abundance and rainfall or physical–chemical parameters,	June 2011 to July 2012 (> 1 year)	Kadir et al., 2020

No.	Fish or/and Shellfish (Common name, Scientific name)*	Climate change issue	Area and location	Significant finding(s) and Discussion	Sampling period	References
19.	Fish (Everrett's dumpis, *Nematabramis everetti*)	Drought	Streams, southeastern Sabah	Distance moved by *N. everetti* was significantly lower during and after the drought, compared to before the drought, as they became more difficult to catch during the drought.	April–June 2013 until February–July 2017 (4 years)	Wilkinson et al., 2019
20.	Fish (Indian mackerel, *Rastrelliger kanagurta*)	Temperature	Open sea, east coast Peninsular Malaysia***	Sea surface temperature (29–30 °C) influenced fish aggregations. Temperature indicates the changes and variability of the monsoon and is a vital variable for detecting aggregation areas of *R. kanagurta*, which are sensitive to changes in temperature.	2008 until 2010 (2 years)	Yeny-Nadira et al., 2019
21.	Shellfish (Banana shrimp, *Penaeus merguiensis*)	Temperature, salinity, rainfall	Mangrove area, Kuala Sepetang, Perak	No significant climate change factors related to shrimp emigration. Pattern of rainfall in the Matang area may lead to a relatively constant emigration of larger juvenile prawns to estuarine areas.	January 1991 until October 1995 (4 years)	Adnan et al., 2002

Table 6. (Continued)

No.	Fish or/and Shellfish (Common name, Scientific name)[*]	Climate change issue	Area and location	Significant finding(s) and Discussion	Sampling period	References
22.	Fish (20 species[***])	Anthropogenic activities, temperature, salinity, rainfall	Peat swamp, Tanjong Karang, Selangor	Water temperatures were associated with fish occurrences. This was due to the fact that the temperature is able to modulate physiological activities and induce stress in fish.	June 2015 (dry month), October 2015 (relatively high rainfall month) and January 2016 (moderately dry month) (< 1 year)	Sule et al., 2018

[*] Common name was listed where available, and literature with ≥ 10 is listed in the Appendix 1 (where available – family name not listed in the study, only species scientific.

[**] Based on the author's response to the corresponding author's email.

[***] Please refer to Appendix 1.

Table 7. Types of data collected related to the impacts of climate change on aquatic fish and shellfish in Malaysia

No.	Types of data	Availability of data	Details of main data found	Conclusion	References
1.	Chemical changes (Pollution – Contamination of heavy metals)	Level of mercury (Hg) and methyl mercury (MeHg) caused by agricultural runoff	The Hg concentrations were in the range of 65.13–102.12 μg/kg and 75.75–106.10 μg/kg for the dry and rainy seasons, respectively. Meanwhile, for MeHg, concentrations in the dry and rainy seasons were in the range of 4.35–6.26 μg/kg and 5.42–6.46 μg/kg, respectively.	Total Hg and MeHg below the limit set by Malaysia Food Act; however, it is safe to consume due to the low ingestion dose rate and health risk index (HRI) for human health.	Abdullah et al., 2015a
2.	Chemical changes (Pollution – Appearances of natural radioactivity)	Accumulation of radionuclides (thorium, uranium, radium-226, radium-228 and potassium-40) caused by the anthropogenic activities	Potassium in the dry and wet seasons was in the range of 1.64–2.23% and 1.12–1.84%, respectively. Thorium concentrations were in the range of 3.62–5.08 mg/kg and 3.78–4.46 mg/kg for the dry and wet seasons, respectively. Uranium was in the range of 1.06–1.26 mg/kg and 0.85–1.17 mg/kg for the dry and rainy seasons, respectively.	Living habitat affected the basis for accumulation of radionuclides in the sampled animals.	Abdullah et al., 2015b
3.	Chemical changes (Pollution – Appearances of phenyltins)	Distribution of organotin and biocides, caused by agricultural runoff	*P. viridis* and *P. nobilis* contained the highest monophenyltin (MPT) concentrations: 161.91 and 43.9 μg/kg, respectively.	Contaminants of organotin and biocides indicated the compounds were discharged into marine environment through agricultural activities.	Mukhtar et al., 2020

Table 7. (Continued)

No.	Types of data	Availability of data	Details of main data found	Conclusion	References
4.	Physiological changes (Infection – Appearances of parasitism)	Total number of parasites, associated with the heavy rainfall pattern	The highest intensity of parasite infection was observed in the rainy season (n = 5014), while the lowest number of infections was in the dry season (n = 767).	The infection levels were higher during the rainy season, which was characterized by low surface seawater temperature, low salinity, and heavy rainfall.	Zakariah et al., 2019
5.	Physiological changes (Growth increment)	Shell increment width change when increasing the level of seawater temperature and salinity	Maximum width of the cockle (65.08 ± 3.01 mm) was observed at the optimum temperature (30 °C).	Significant differences in the growth rate of cockle shells were mainly controlled by temperature and less by salinity.	Mirzaei and Hwai, 2015
6.	Physiological changes (Mortality)	Relationship between *Anadara granosa* mass mortality and heavy rainfall	Heavy rainfall between mid-February (over 30–50 mm/day) caused 30% higher mortality rate in cockles.	Blood cockles which live near the estuary were considered to have been adversely impacted by flooding (caused by heavy rainfall) from nearby rivers.	Yurimoto et al., 2014

No.	Types of data	Availability of data	Details of main data found	Conclusion	References
7.	Physiological changes (Shell growth and gonad development)	Correlations between increment widths, gonad development and environmental factors (temperature and salinity)	Positive correlation between salinity and increment in width and main spawning season (gametogenic activity occurred) coincided with high seawater temperatures (30–33 °C) and considerable fluctuations in seawater salinity (14 ppt).	Growth breaks in *A. granosa* shell cross-sections are formed by shell growth interruption. These growth breaks arise from various environmental stressors, including temperature or salinity shocks.	Mirzaei et al., 2017
8.	Behavioural changes (Spatial distribution pattern)	Sediment properties and key environmental parameters responsible for distribution of fiddler crabs	The Fiddler crab (genus *Uca*) distribution depends on the species. pH was an important factor in determining the *U. forcipata* habitat, with an optimum range of 7.22 ± 0.11. Temperature significantly affected 13 different factors of environmental changes and crab burrow density, as well as sediment properties, which impacted their distribution.	Sediment heterogeneity and environmental parameters responsible for changes in crab community composition.	Mokhtari et al., 2015
9.	Behavioural changes (Species interaction and distribution)	Salinity influenced the mud lobster species distribution	The Mud lobster (genus *Thalassina*) distribution depends on the species. Salinity (between 21.50 to 24.2 ppt) was an important factor affecting *T. kelanang* species distribution in the mangrove area.	*T. kelanang* is less tolerant of large salinity fluctuations, compared to other species (e.g., *T. anomala* and *T. gracilis*).	Moh et al., 2015

Table 7. (Continued)

No.	Types of data	Availability of data	Details of main data found	Conclusion	References
10.	Behavioural changes (Density and distribution)	Total number of fish and its relationship water quality at each sampling station	The two highest total fish counts captured per sampling effort unit were at the station with the highest water temperature (30.2 °C) and dissolved oxygen (2.12–6.73).	Non-native fish species are more resilient to the pollution level in the agricultural runoff (70%).	Yusof et al., 2018
11.	Behavioural changes (Diversity and group)	Total fish abundance associated with environmental variables (conductivity & chemical oxygen demand; COD)	In the canonical correspondence analysis (CCA) on the effect of environmental variables on fish assemblages, the longest CCA vector was recorded for water conductivity, followed by COD.	The importance of COD in characterizing fish assemblages in this upper river habitat.	Ikhwanuddin et al., 2016
12.	Behavioural changes (Diversity and distribution)	Peninsular Malaysia freshwater mussel distribution records, native and non-native species associated with water quality parameters	*Sinanodonta woodiana* was the species that is most tolerant to acidic conditions; other species were associated with high silicate levels.	Several mussel species were threatened by human-induced eutrophication and acidification.	Zieritz et al., 2016
13.	Behavioural changes (Larval abundance)	Correlation and analysis between fish larval abundance and environmental variables	*Pomacentridae* sp. Only found at one station with low total suspended solids, low pH, and low salinity. *M. argentus* did not appear at one station with high salinity.	Weak correlation between larval fish abundance and water quality parameters through CCA.	Ara et al., 2016

No.	Types of data	Availability of data	Details of main data found	Conclusion	References
14.	Behavioural changes (Species richness)	Correlation analysis between fish species richness and extensive anthropogenic activities (agriculture, plantation)	Bornean Spotted Barb (*Barbodes sealei*) accounted for around 21% of the total fish species and, surprisingly, sixteen species*** were only detected in a single land-use type, most of which were also rare.	Streams in all disturbed land-uses for agricultural plantation suffer losses in species richness.	Wilkinson et al., 2018
15.	Behavioural changes (Larval abundance and diversity)	Correlation analysis between fish species assemblage and environmental variables	Gobiidae was the most abundant family, which contributed to 80.68% of the total abundance.	Weak relationship between larval abundance and water quality (i.e., environmental variables) through multi-regression analysis.	Ara et al., 2020
16.	Behavioural changes (Larval assemblage)	Correlation analysis between fish larval abundance and environmental factors	A total of 15 and 17 families were recorded in mangrove and offshore waters, respectively.	Salinity appeared to be the most significant factor influencing the distribution and abundance of most larval fish.	Ooi and Chong, 2011
17.	Behavioural changes (Distribution and habitat preference)	Distribution of species with same genus at different salinity level	Distribution and habitat separation of the two species (*O. javanicus* and *O. dancena*) along river and mangrove areas.	Both species were found in all sampling station, but *O. javanicus* is highly adaptable to full-strength seawater.	Yusof et al., 2012

Table 7. (Continued)

No.	Types of data	Availability of data	Details of main data found	Conclusion	References
18.	Behavioural changes (Abundance)	Correlation and analysis between fish abundance and rainfall and physico-chemical parameters	Salinity was the most important variable in explaining variation in fish species composition, with *Epinephelus coioides, Lutjanus russellii, Toxotes jaculatrix, Yarica hyalosoma,* and *Dichotomyctere nigroviridis* highly abundant under low salinity.	No significant difference between fish abundance, rainfall, and other environmental factors in the regression analysis.	Kadir et al., 2020
19.	Behavioural changes (Distance moved)	Impact of drought (before, during and after) and their interaction with fish density and movement	Interaction of *Nematabramis everetti* with drought, in terms of fish density and distance moved, with density ranging from 138 ± 37 to 2774 ± 1004 fish km^{-1} across the 11 streams.	Fish catchability was reduced during drought, indicating that drought may exert a negative impact on local communities of the fish species.	Wilkinson et al., 2019
20.	Behavioural changes (Abundance)	Correlation between fish abundance and variability in oceanographic conditions of temperature	Environmental variables of temperature (29–31 °C) influenced *R. kanagurta* in terms of aggregation areas.	High distribution of Indian mackerel was related to sea surface temperature.	Yeny-Nadira et al., 2019

No.	Types of data	Availability of data	Details of main data found	Conclusion	References
21.	Behavioural changes (Emigration pattern)	Environmental factors influence the emigration of larger juveniles from the estuary	Banana shrimp remained at the mouth for a further 8–12 weeks before emigrating from the river.	The juveniles appeared to start emigrating from the estuary at a size of 12–17 mm carapace length, and were not affected by all environmental factors.	Adnan et al., 2002
22.	Behavioural changes (Occurrences, migration and distribution	Environmental factors and anthropogenic activities influence fish occurrences, migration and distribution	Only 15 fish species were found to be related to water physico-chemical parameters in peat swamps, while temperature was related to the occurrence of most fish species.	Progressive decrease in fish diversity recorded due to anthropogenic activities. Eight species were influenced by water temperature.	Sule et al., 2018

*** Please refer to the Appendix 1.

An increase in salinity will have an impact on the energy-demanding physiological activities of organisms, such as reproduction. Higher gametogenic activity was observed with significant changes in seawater salinity (14 ppt to 30 ppt) (Mirzaei et al., 2017). Furthermore, one review found that osmotic shock caused by heavy rainfall contributed to cockle mass mortality (Yurimoto et al., 2014). High precipitation also caused salinity fluctuations, which have an impact on the osmoregulation of marine bivalves (Patrick et al., 2019). Salinity below 15 ppt caused 100% mortality rate in marine cockles (*Cerastoderma edule*) as well as valve-closing reactions and depressed physiological activity (Peteiro et al., 2018). It was demonstrated that *C. edule* shut its valve to isolate the organism from salinity fluctuations, triggering anaerobic metabolism. Prolonged isolation led to longer exposures exceeding their critical tolerance, resulting in mass mortality events (Peteiro et al., 2018). Zakariah et al., (2019) discovered a significant intensity of protozoan parasites associated with heavy rainfall in their study (i.e., high precipitation). This could be due to the fact that rainfall is a major driver of seasonal variation in parasitism, with heavy rainfall having direct effects on parasite loads, thus affecting parasite survival and movement in the environment (Shearer and Ezenwa, 2020). Climate change is expected to have a negative impact on the physiological changes of fish and shellfish species, particularly in terms of growth, survival, and development, due to the increased frequency of unpredictable weather forecasts.

Behavioural Changes

Several studies have found that climate change causes significant behavioural changes in specific organisms (Ooi and Chong, 2011; Yusof et al., 2012; Ara et al., 2016; Ikhwanuddin et al., 2016; Sule et al., 2018; Wilkinson et al., 2018; Wilkinson et al., 2019; Yusof et al., 2018; Yuslan et al., 2021; Yeny-Nadira et al., 2019; Ara et al., 2020; Kadir et al., 2020) and shellfish (Adnan et al., 2002; Moh et al., 2015; Mokhtari et al., 2015; Zieritz et al., 2016). Climate change has been linked to behavioural changes related to environmental factors, such as distribution pattern, abundance, habitat preferences, distance movement, and emigration pattern. Heath et al., (2012) have provided evidence for such a pattern, as they found climate change to have a great impact on the behaviour of fish and shellfish from the United Kingdom and Ireland.

Temperature and salinity were found to influence the habitat determination and distribution pattern of fiddler crab (genus *Uca*) and mud lobster (genus *Thalassina*) (Mokhtari et al., 2015; Moh et al., 2015). Low water temperature determined the poleward range limits of the South

American fiddler crab species *Leptuca cumulanta* (Grande et al., 2021). In this context, understanding the thermal tolerance and biology of ectothermic animals is critical to gaining a better understanding of their distribution pattern (Heath et al., 2012; Syafaat et al., 2021). Some crustacean species may have varying thermal tolerance ranges and preferences, which have direct impacts on their movement and distribution (Azra et al., 2018). Thermal biological data related to the behavioural pattern of a species are critical for better understanding its interaction with changes in seawater/freshwater temperature. This will help future research focused on predicting changes in geographic distribution caused by climate change (Bricknell et al., 2021). Fiddler crabs are considered a keystone species in intertidal environments, due to their ability to survive in conditions with high salinity fluctuations (between 0.5 ppt and 30 ppt). This expectedly leads to their frequent movement to more tolerable habitats, hence presenting varying abundance and distribution across the ecosystem. The difference in osmoregulation capacities in crustacean species across the embryonic, juvenile, and adult stages might be species- and stage-specific. Gills and antennal glands may play important roles in salinity adaptation for many shellfish species that live in high salinity environments, such as brackish water and intertidal zones (Thabet et al., 2017). Zieritz et al., (2016) discovered that water quality requirements had a significant impact on the abundance of freshwater mussels in their respective habitats. Surprisingly, heavy rainfall caused by climate change had no effect on the emigration pattern of banana shrimp (Adnan et al., 2002). Other findings, however, revealed a strong correlation and relationship between climate change and environmental factors such as dissolved oxygen, temperature, and fish abundance (Ooi and Chong, 2011; Yusof et al., 2012; Ara et al., 2016; Ikhwanuddin et al., 2016; Sule et al., 2018; Yusof et al., 2018; Wilkinson et al., 2019; Yeny-Nadira et al., 2019; Ara et al., 2020). Temperature has an effect on the chemical changes in water, including the composition and pressure of dissolved gases. Previous research in fish sampling sites demonstrated an inverse correlation between temperature and solubility of gases in water (dissolved oxygen). This accounted for the highest number of fish recorded in the vicinity of the sampling sites where the lowest water temperature was found. Equal volumes of freshwater at the same temperature and pressure will hold more dissolved gases than brackish or marine waters. Species diversity and the spatiotemporal abundance of fish communities are affected by environmental factors such as climate, temperature, salinity, and oxidation levels (Prakash, 2021). Climate change will influence the behaviour of fish and shellfish by altering their movement and abundance, potentially affecting the biodiversity in their new preferred habitats. In the current chapter, we noted a lack of research focused on the impacts of climate change on freshwater habitats associated with shellfish species. Furthermore, there is also a scarcity of research on the impacts of climate change on fish species in natural marine habitats. This could be attributed to the fact that studies on marine species have mainly been focused on the laboratory examination of cultured fish or shellfish species (De et al., 2019; Noor et al., 2019).

Conclusion

Limitations

In this chapter, we only considered scholarly articles retrieved from the WoS core collection and Scopus. Other sources, such as PubMed, Google Scholar, or grey literature, were not included. Furthermore, the articles were only focused on studies within Malaysia (or conducted in Malaysia by foreign researchers), albeit in the different regions with varying climate impacts. It is possible that cultural perspectives or beliefs about the effects of climate change may differ in other countries.

Implications for Policy/Management

This is the first comprehensive review on the impacts of climate change on aquatic animals in Malaysia, which parallels reports of negative consequences in other countries. These negative impacts are very challenging to the environment and can be categorised into three general effects: (i) Chemical, (ii) physiological, and (iii) behavioural. Thus, there is an urgent need for policymakers to develop mitigation strategies focused on improving the sustainability and resilience of aquatic animals against climate change-related effects.

It is recommended that policymakers focus on protecting "ecological engineers," such as corals, in order to improve the resilience of other species dependent on them, consequently improving the health of aquatic animals in Malaysia. One of the most popular mitigation solutions is the establishment of large marine protected areas, in which fisheries activities are controlled; for example, in terms of the allowed catch and size.

It is recommended that the use of advanced equipment and integrated approaches is increased, such as the implementation of integrated coastal zone management, in order to reduce reliance on top-down approaches and focus on bottom-up policy prevention, ultimately allowing for increased species diversity, species with high added value, and survival rates. We recommend additional research or analysis on the topic of climate change impacts on aquatic animals, in order to address specific knowledge gaps. One potential research area to develop is the establishment of aquatic ecosystems and fisheries for fish and shellfish species that are less sensitive to climate change. Assisting aquatic ecosystems and communities to prepare for change and become more resilient to the effects of climate change through specific and structured programmes is also strongly encouraged.

References

Abdullah, A., Hamzah, Z., Saat, A., Wood, A. K. & Alias, M. (2015a). Accumulation of mercury (Hg) and methyl mercury (MeHg) concentrations in selected marine biota from Manjung coastal area. Malays. *J. Anal. Sci.* 19: 669 – 678. http://mjas.analis.com.my/wp-content/uploads/2018/11/Anisa_19_4_4.pdf.

Abdullah, A., Hamzah, Z., Saat, A., Wood, A. K. & Alias, M. (2015b). Accumulation of radionuclides in selected marine biota from Manjung coastal area. *AIP Conf. Proc.* 1659: 050009. https://doi.org/10.1063/1.4916879.

Adnan, N. A., Loneragan, N. R. & Connolly, R. M. (2002). Variability of, and the influence of environmental factors on, the recruitment of postlarval and juvenile *Penaeus merguiensis* in the Matang mangroves of Malaysia. *Mar. Biol.* 141: 241-251. https://doi.org/10.1007/s00227-002-0837-2.

Aghamohammadi, N. & Isahak, M. (2017). *Climate change and air pollution in Malaysia. In* Climate change and air pollution the impact on human health in developed and developing countries. Eds Akhtar, R. and Palagiano, C. Springer Climate. Springer, Cham. Pp: 241-254. https://link.springer.com/content/pdf/10.1007%2F978-3-319-61346-8.pdf.

Ali, J. M., D'Souza, D. L., Schwarz, K., Allmon, L. G., Singh, R. P., Snow, D. D., Bartelt-Hunt, S. L. & Kolol, A. S. (2018). Response and recovery of fathead minnows (*Pimephales promelas*) following early life exposure to water and sediment found within agricultural runoff from the Elkhorn River, Nebraska, USA. *Sci. Total Environ.* 618: 1371-1381. https://doi.org/10.1016/j.scitotenv.2017.09.259.

Ara, R., Arshad, A., Amin, S. M. N., Idris, M. H., Gaffar, M. A. & Romano, N. (2016). Influence of habitat structure and environmental variables on larval fish assemblage in the Johor Strait, Malaysia. *J. Environ. Biol.* 37: 745-754. http://www.jeb.co.in/journal_issues/201607_jul16_spl/paper_14.pdf.

Ara, R., Arshad, A., Amin, S. M. N., Yusoff, F. M. & Romano, N. (2020). Larval fish assemblage, diversity and habitat ecology in the Matang Mangrove Ecosystem, Perak, Malaysia. *J. Environ. Biol.* 41: 1316-1325. http://www.jeb.co.in/journal_issues/202009_sep20_spl/paper_26.pdf.

Azra, M. N., Chen, J. C., Ikhwanuddin, M. & Abol-Munafi, A. B. (2018). Thermal tolerance and locomotor activity of blue swimmer crab *Portunus pelagicus* instar reared at different temperatures. *J. Therm. Biol.* 74: 234-240. https://doi.org/10.1016/j.jtherbio.2018.04.002.

Azra, M. N., Aaqillah-Amr, M. A., Ikhwanuddin, M., Ma, H., Waiho, K., Ostrensky, A., Tavares, C. P. D. S. & Abol-Munafi, A. B. (2020). Effects of climate-induced water temperature changes on the life history of brachyuran crabs. *Rev. Aquac.* 12: 1211-1216. https://doi.org/10.1111/raq.12380.

Azzeri, A., Ching, G. H., Jaafar, H., Noor, M. I. M., Razi, N. A., Then, A. Y. H., Suhaimi, J., Kari, D. & Dahlui, M. (2020). A review of published literature regarding health issues of coastal communities in Sabah, Malaysia. *Int. J. Environ. Res. Public Health.* 17:5. https://doi.org/10.3390/ijerph17051533.

Belur, J., Tompson, L., Thornton, A., & Simon, M. (2021). Interrater reliability in systematic review methodology: exploring variation in coder decision-making. *Sociol. Methods Res.* 50:2: 837-865. https://doi.org/10.1177/0049124118799372.

Bricknell, I. R., Birkel, S. D., Brawley, S. H., Kirk, T. V., Hamlin, H. J., Capistrant-Fossa, K., Huguenard, K., Walsum, G. P. V., Liu, Z. L., Zhu, L. H., Grebe, G., Taccardi, E., Miller, M., Preziosi, B. M., Duffy, K., Byron, C. J., Quigley, C. T. C., Bowden, T. J., Brady, D., Beal, B. F., Sappati, P. K., Johnson, T. R. & Moeykens, S. (2021). Resilience of cold-water aquaculture: a review of likely scenarios as climate changes in the Gulf of Maine. *Rev. Aquac.* 13: 460-503. https://doi.org/10.1111/raq.12483.

Burrows, M. T., Schoeman, D. S., Buckley, L. B., Moore, P., Poloczanska, E. S., Brander, K. M., Brown, C., Bruno, J. F., Duarte, C. M., Halpern, B. S. & Holding J. (2011). The pace of shifting climate in marine and terrestrial ecosystems. *Science.* 334(6056):652-655. https://doi.org/10.1126/science.1210288.

Charles, S. P., Kominoski, J. S., Armitage, A. R., Guo, H., Weaver, C. A. & Pennings, S. C. (2020). Quantifying how changing mangrove cover affects ecosystem carbon storage in coastal wetlands. *Ecology.* 101: e02916. https://doi.org/10.1002/ecy.2916.

De, M., Mazlan, A. G., Noor, N. M. Cob, Z. C., Bakar, Y. & Das, S. K. (2019). Effects of water temperature and diet on blood parameters and stress levels in hybrid grouper (*Epinephelus fuscoguttatus* ♀ × *E. lanceolatus* ♂) juveniles. *Aquacult. Rep.* 15: 100219. https://doi.org/10.1016/j.aqrep.2019.100219.

DoS. (2020). Department of Statistics Malaysia (DoS), Agriculture: Selected Agricultural Indicators, Malaysia, 2020. https://www.dosm.gov.my/v1/index.php?r=column/ctwoByCat&parent_id=45&menu_id=Z0VTZGU1UHBUT1VJMFlpaXRRR0xpdz09.

Ercan, A., Mohamad, M. F. B. & Kavvas, M. L. (2012). The impact of climate change on sea level rise at Peninsular Malaysia and Sabah–Sarawak. *Hydrol. Process.* 27: 367-377. https://doi.org/10.1002/hyp.9232.

Firdaus, R. B. R., Tan, M. L., Rahmat, S. R. & Gunaratne, M. S. (2020). Paddy, rice and food security in Malaysia: A review of climate change impacts. *Cogent Soc. Sci.* 6: 1818373. https://doi.org/10.1080/23311886.2020.1818373.

Gilbert, P. J., Taylor, S., Cooke, D. A., Deary, M. E. & Jeffries, M. J. (2021). Quantifying organic carbon storage in temperate pond sediments. *J. Environ. Manage*. 280: 111698. https://doi.org/10.1016/j.jenvman.2020.111698.

Gopalakrishnan, T., Hasan, M. K., Haque, A. T. M. S., Jayasinghe, S. L. & Kumar, L. (2019). Sustainability of coastal agriculture under climate change. *Sustainability*. 11: 7200. http://dx.doi.org/10.3390/su11247200.

Goulas, A., Belhadi, D., Descamps, A., Andremont, A., Benoit, P., Courtois, S., Dagot, C., Grall, N., Makowski, D., Nazaret, S., Nélieu, S., Patureau, D., Petit, F., Roose-Amsaleg, C., Vittecoq, M., Livoreil, B., & Laouénan, C. (2020). How effective are strategies to control the dissemination of antibiotic resistance in the environment? A systematic review. *Environ. Evid.* 9:1. https://doi.org/10.1186/s13750-020-0187-x.

Grande, F. R. D., Arakaki, J. Y., Marochi, M. & Coasta, T. M. (2021). Cold water temperatures define the poleward range limits of south American fiddler crabs. *Estuar. Coast. Shelf Sci*. 260: 107494. https://doi.org/10.1016/j.ecss.2021.107494.

Griffiths, G. M., Chambers, L. E., Haylock, M. R., Manton, M. J., Nicholls, N., Baek, H. J., Choi, Y., Della-Marta, P. M., Gosai, A., Iga, N., Lata, R., Laurent, V., Maitrepierre, L., Nakamigawa, H., Ouprasitwong, N., Solofa, D., Tahani, L., Thuy, D. T., Tibig, L., Vediapan, K. & Zhai, P. (2005). Change in mean temperature as a predictor of extreme temperature change in the Asia-Pacific region. *Int. J. Climatol*. 25: 1301-1330. https://doi.org/10.1002/joc.1194.

Haddaway, N. R., Macura, B., Whaley, P., & Pullin, A. S. 2017. *ROSES for Systematic Review Reports*. Version 1.0. https://doi.org/10.6084/m9.figshare.5897272.

Häkkilä, M., Johansson, A., Sandgren, T., Uusitalo, A., Mönkkönen, M., Puttonen, P., & Savilaakso, S. (2021). Are small protected habitat patches within boreal production forests effective in conserving species richness, abundance and community composition? A systematic review. *Environ. Evid.* 10:1. https://doi.org/10.1186/s13750-020-00216-6.

Heath, M. R., Neat, F. C., Pinnegar, J. K., Reid, D. G., Sims, D. W. & Wright, P. J. (2012). Review of climate change impacts on marine fish and shellfish around the UK and Ireland. *Aquat. Conserv*. 22: 337-367. https://doi.org/10.1002/aqc.2244.

Ikhwanuddin, M. E. M., Amal, M. N. A., Shohaimi, S., Hasan, H. H. & Jamil, N. R. (2016). Environmental influences on fish assemblages of the upper Sungai Pelus, Kuala Kangsar, Perak, Malaysia. *Sains Malays*. 45: 1487-1495. http://www.ukm.my/jsm/pdf_files/SM-PDF-45-10-2016/09%20Mat%20Esa%20Mohd%20Ikhwanuddin.pdf.

Joni, A. A. M., Mohamat-Yusuff, F., Noor, N. A. M., Mohamed, K. N., Ash'aari, Z. H., Kusin, F. M., Zulkeflee, Z., Zulkifli, S. Z., Ismail, A. & Arshad, A. (2021). Baseline distribution and sources of selected agricultural runoff in the bottom water of an active cockle farming area, Bagan Pasir, Perak, Malaysia. *Mar. Pollut. Bull*. 167: 112276. https://doi.org/10.1016/j.marpolbul.2021.112276.

Kadir, S. T. S. A., Manaf, N. A., Ali, A. N., Wail, N. F. C., Bidai, J. A., Ahmad, H., Abdullah, C. M. K. A. C., Mohamad, M. N., Ahmad, M. F., Ahmad, A., Bachok, Z., Motomura, H., Husain, M. L., Ambak, M. A. & Ghaffar, M.A. (2020). Physical-chemical parameters in relation to fish assemblages in the East Coast of Peninsular Malaysia. *Malays. Appl. Biol. J.* 49: 193-199. https://www.mabjournal.com/images/49_4_December_2020/49_04_26.pdf.

Matias, D. M. S., Kim, H., Anticamara, J. A., Matias, K. M. S. & Tibig, L. V. (2021). Emerging response options and scenarios of slow onset events related to climate change in Southeast Asia. *Curr. Opin. Environ. Sustain.* 50: 175-184. https://doi.org/10.1016/j.cosust.2021.04.004.

Mirzaei, M. R., Yasin, Z. & Hwai, A. T. S. (2014). Periodicity and shell microgrowth pattern formation in intertidal and subtidal areas using shell cross sections of the blood cockle, *Anadara granosa*. *Egypt. J. Aquat. Res.* 40: 459-468. https://doi.org/10.1016/j.ejar.2014.11.008.

Mirzaei, M. R. & Hwai, A. T. S. (2015). Assessing cockle shells (*Anadara granosa*) for reconstruction sub-daily environmental parameters: Implication for paleoclimate studies. *Hist. Biol.* 28: 896-906. https://doi.org/10.1080/08912963.2015.1052806.

Mirzaei, M. R., Hwai, A. T. S. & Khalil, M. (2017). Temporal variation in shell growth rate of cockle *Anadara granosa* in relation with its reproductive cycle. *J. Shellfish Res.* 36: 69-78. http://dx.doi.org/10.2983/035.036.0109.

Moh, H. H., Chong, V. C. & Sasekumar, A. (2015). Distribution and burrow morphology of three sympatric species of *Thalassina* mud lobsters in relation to environmental parameters on a Malayan mangrove shore. *J. Sea Res.* 95: 75-83. https://doi.org/10.1016/j.seares.2014.10.006.

Mokhtari, M., Ghaffar, M. A., Usup, G. & Cob, Z. C. (2015). Determination of key environmental factors responsible for distribution patterns of fiddler crabs in a tropical mangrove ecosystem. *PloS One* 10(1): e0117467. https://doi.org/10.1371/journal.pone.0117467.

Mukhtar, A., Zulkifli, S. Z., Mohamat-Yusuff, F., Harino, H. & Ismail, A. (2020). Distribution of biocides in selected marine organisms from South of Johor, Malaysia. *Reg. Stud. Mar. Sci.* 38: 101384. https://doi.org/10.1016/j.rsma.2020.101384.

Nagelkerken, I., Blaber, S. J. M., Bouillon, S., Green, P., Haywood, M., Kirton, L. G., Meynecke, J. O., Pawlik, J., Penrose, H. M., Sasekumar, A. & Somerfield, P. J. (2008). The habitat function of mangroves for terrestrial and marine fauna: A review. *Aquat. Bot.* 89: 155-185. https://doi.org/10.1016/j.aquabot.2007.12.007.

Noor, N. M., De, M., Iskandar, A., Keng, W. L., Cob, Z. C., Ghaffar, M. A. & Das, S. K. (2019). Effects of elevated carbon dioxide on the growth and welfare of Juvenile tiger grouper (*Epinephelus fuscoguttatus*) × giant grouper (*E. lanceolatus*) hybrid. *Aquaculture*. 513: 734448. https://doi.org/10.1016/j.aquaculture.2019.734448.

Ooi, A. L. & Chong, V. C. 2011. Larval fish assemblages in a tropical mangrove estuary and adjacent coastal waters: Offshore-inshore flux of marine and estuarine species. *Cont. Shelf Res.* 31: 1599-1610. https://doi.org/10.1016/j.csr.2011.06.016.

Ong, E. Z., Briffa, M., Moens, T. & Colen, C. V. (2017). Physiological responses to ocean acidification and warming synergistically reduce condition of the common cockle *Cerastoderma edule*. *Mar. Environ. Res.* 130: 38-47. https://doi.org/10.1016/j.marenvres.2017.07.001.

Park, C. U. & Kim, H. J. (2015). Measurement of inter-rater reliability in systematic review. *Hanyang Med. Rev.* 35:1:44. https://doi.org/10.7599/hmr.2015.35.1.44.

Patrick, C. J., McGarvey, D. J., Larson, J. H., Cross, W. F., Allen, D. C., Benke, A. C., Brey, T., Huryn, A. D., Jones, J., Murphy, C. A., Ruffing, C., Saffarinia, P., Whiles, M. R., Wallace, J. B. & Woodward, G. (2019). Precipitation and temperature drive continental-scale patterns in stream invertebrate production. *Sci. Adv.* 5: eaav2348. https://doi.org/10.1126/sciadv.aav2348.

Peteiro, L. G., Woodin, S. A., Wethey, D. S., Costas-Costas, D., Martinez-Casal, A., Olabarria, C. & Vazquez, E. (2018). Responses to salinity stress in bivalves: Evidence of ontogenetic changes in energetic physiology on *Cerastoderma edule*. *Sci. Rep.* 8: 8329. https://doi.org/10.1038/s41598-018-26706-9.

Prakash, C. (2021). Aquatic health and aquaculture. CRC Press, New York. Pp 204. ISBN: 978-1-003-11152-8 (e-book).

Rosli, N. A. M. & Zain, K. M. (2016). Preliminary assessment on autecological studies of Beardless Barb, *Cyclocheilichthys apogon* (Valenciennes, 1842) from Muda Reservoir of Kedah, Malaysia. *Trop. Life Sci. Res.* 27: 63-69. https://dx.doi.org/10.21315%2Ftlsr2016.27.3.9.

Shearer, C. L. & Ezenwa, V. O. (2020). Rainfall as a driver of seasonality in parasitism. *Int. J. Parasitol. Parasites Wildl.* 12: 8-12. https://doi.org/10.1016/j.ijppaw.2020.04.004.

Silvius, M. & Schrier A. (2016) *Climate Regulation: South East Asian Peat Swamps*. In: Finlayson C. et al. (eds) The Wetland Book. Springer, Dordrecht. https://doi.org/10.1007/978-94-007-6172-8_237-3.

Sule, H. A., Ismail, A., Amal, M. N. A., Zulkifli, S. Z., Roseli, M. F. A. M. & Shohaimi, S. (2018). Water quality influences on fish occurrence in peat swamp forest and its converted areas in north Selangor, Malaysia. *Sains Malays.* 47: 2589-2600. http://dx.doi.org/10.17576/jsm-2018-4711-01.

Syafaat, M. N., Azra, M. N., Mohamad, F., Che-Ismail, C. Z., Amin-Safwan, A., Asmat-Ullah, A., Syahnon, M., Azmie, G., Abol-Munafi, A. B., Ma, H. & Ikhwanuddin, M. (2021). Thermal tolerance and physiological changes of mud crab, *Scylla paramamosain* crablet at different water temperatures. *Animals.* 11: 1146. https://doi.org/10.3390/ani11041146.

Tan, M. L., Ibrahim, A. L., Cracknell, A. P. & Yusop, Z. (2017). Changes in precipitation extremes over the Kelantan River Basin, Malaysia. *Int. J. Climatol.* 37: 3780-3797. https://doi.org/10.1002/joc.4952.

Tan, M. L., Juneng, L., Tangang, F. T., Chung, J. X. & Firdaus, R. B. R. (2021). Changes in temperature extremes and their relationship with ENSO in Malaysia from 1985 to 2018. *Int. J. Climatol.* 41: E2564-E2580. https://doi.org/10.1002/joc.6864.

Tang, K. H. D. (2019). Climate change in Malaysia: Trends, contributors, impacts, mitigation and adaptations. *Sci. Total Environ.* 650: 1858-1871. https://doi.org/10.1016/j.scitotenv.2018.09.316.

Tangang, F. T., Juneg, L., Salimun, E., Sei, K. M., Le, L. J. & Muhamad, H. (2012). Climate change and variability over Malaysia: Gaps in science and research information. *Sains Malays.* 41: 1355-1366. http://www.ukm.my/jsm/pdf_files/SM-PDF-41-11-2012/05%20Fredolin%20T%20Tangang.pdf.

Tangang, F. T., Juneng, L., Salimun, E., Sei, K. M. S., Le, L. J. & Muhamad, H. (2012). Climate change and variability over Malaysia: Gaps in science and research information. *Sains Malays.* 41: 1355-1366. http://www.ukm.my/jsm/pdf_files/SM-PDF-41-11-2012/05%20Fredolin%20T%20Tangang.pdf.

Thabet, R., Ayadi, H., Koken, M. & Leignel, V. (2017). Homeostatic responses of crustaceans to salinity changes. *Hydrobiologia*. 799: 1-20. https://doi.org/10.1007/s10750-017-3232-1.

Walsh, R. P., Ellison, S., Los, S. O., Bidin, K., Sayer, A. M. & Tussin, A. M. (2013). Changes in large rainstorm magnitude–frequency over the last century in Sabah, Malaysian Borneo and their geomorphological implications. *Holocene*. 23: 1824-1840. https://doi.org/10.1177%2F0959683613507791.

Wilkinson, C. L., Yeo, D. C. J., Hui, T. H., Fikri, A. H. & Ewers, R. M. 2018. Land-use change is associated with a significant loss of freshwater fish species and functional richness in Sabah, Malaysia. *Biol. Conserv*. 222: 164-171. https://doi.org/10.1016/j.biocon.2018.04.004.

Wilkinson, C., Yeo, D. C. J., Tan, H. H, Fikri, A. H. & Ewers, R. M. (2019). Resilience of tropical, freshwater fish (*Nematabramis everetti*) populations to severe drought over a land-use gradient in Borneo. *Environ. Res. Lett*. 41: 045008. https://doi.org/10.1088/1748-9326/ab0128.

Yeny-Nadira, K., Mustapha, M. A. & Ghaffar, M. A. (2019). The Indian mackerel aggregation areas in relation to their oceanographic conditions. *Sains Malays*. 48: 2575-2581. http://dx.doi.org/10.17576/jsm-2019-4811-27.

Yurimoto, T. Kassim, F. M., Fuseya, R. & Man, A. 2014. Mass mortality event of the blood cockle, *Anadara granosa*, in aquaculture ground along Selangor coast, Peninsular Malaysia. *Int. Aquat. Res*. 6: 177–186. https://doi.org/10.1007/s40071-014-0077-3.

Yuslan, A., Najuwa, S., Hagiwara, A., Ghaffar, M. A., Suhaimi, H., & Rasdi, N. W. (2021). Production performance of *Moina macrocopa* (Straus 1820) (crustacea, cladocera) cultured in different salinities: the effect on growth, survival, reproduction, and fatty acid composition of the neonates. *Diversity*, *13*(3), 105.https://doi.org/10.3390/d13030105.

Yusof, F. M., Jamil, N. R., Roseli, F. A. M., Malik, N. K. A. & Azmai, M. N. A. (2018). Effects of water quality in the agricultural drainage towards the fish diversity and its abundance in the southwest part of Bagan Datuk District, Perak, Malaysia. *Malay. Nat. J.* 70: 373-385.

Yusof, S., Ismail, A., Koito, T., Kinoshita, M. & Inoue, K. (2012). Occurrence of two closely related ricefishes, Javanese medaka (*Oryzias javanicus*) and Indian medaka (*O. dancena*) at sites with different salinity in Peninsular Malaysia. *Environ. Biol. Fishes*. 93: 43-49. https://doi.org/10.1007/s10641-011-9888-x.

Yusoff, F. M., Shariff, M. & Gopinath, N. (2006). Diversity of Malaysian aquatic ecosystems and resources. *Aquat. Ecosyst. Health Manag*. 9: 119-135. https://doi.org/10.1080/14634980600713315.

Zakariah, M. I., Daud, H. M., Sharma, R. S. K., Ikhwanuddin, M. & Hassan, M. (2019). Distribution patterns of gregarine parasitism of wild marine bivalve, Anadara cornea (reeve, 1844) concerning seasonality and water quality. IOP Conf. Ser.: *Earth Environ. Sci.* 370 012063. https://doi.org/10.1088/1755-1315/370/1/012063.

Zieritz, A., Lopes-Lima, M., Bogan, A. E., Sousa, R., Walton, S., Rahim, K. A. A., Wilson, J. J., Ng, P. Y., Froufe, E. & McGowan, S. 2016. Factors driving changes in freshwater mussel (Bivalvia, Unionida) diversity and distribution in Peninsular Malaysia. *Sci. Total Environ*. 571: 1069-1078. http://dx.doi.org/10.1016/j.scitotenv.2016.07.098.

Appendix 1

No.	Content	References
1.	Common and species name: Climbing perch, *Anabas testudineus*; Striped snakehead, *Channa striata*; Mozambique tilapia, *Oreochromis mossambicus*; Mullet, *Crenimugil* sp.; Croaking gourami, T*richopsis vittata*; Three-spot gouramy, *Trichopodus trichopterus*; Siamese gouramy, *Trichopodus pectoralis*; Java catfish, *Hemibagrus nemurus*; Java catfish, *Hemibagrus nemurus*; Malaysian tarpon, *Megalops* sp.; Goby, *Acentrogobius* sp.	Yusof et al., 2018
2.	*Homaloptera orthogoniata, Homalopteroides tweediei, Barbodes binotatus, Crossocheilus atrilimes, Devario regina, Hampala macrolepidota, Lobocheilos bo, Mystacoleucus marginatus Neolissochilus hexagonolepis, Rasbora vulgaris, Tor tambra, Nemacheilus masyae, Channa gachua, Amblyceps foratum, Batasio fluviatilis, Leiocassis poecilopterus, Mystus cavasius, Clarias leiacanthus, Glyptothorax siamensis, Glyptothorax major, Macrognathus keithi*	Ikhwanuddin et al., 2016
3.	Eleven species were significantly associated with higher COD environment: *H. tweediei, Lobocheilos bo, Crossocheilus atrilimes, Amblyceps foratum, Batasio fluviatilis, Rasbora vulgaris, Mystacoleucus marginatus, Leiocassis poecilopterus, Nemacheilus masyae, M. keithi* and *Clarias leiacanthus* were associated with sites with lower COD, while *Hampala macrolepidota, G. siamensis, Devario regina, Mystus cavasius* and *G. major*	Ikhwanuddin et al., 2016
4.	Freshwater habitats (rivers, streams, canals, rice-paddy run-offs, lakes, reservoirs and ponds), Peninsular Malaysia state: Perlis, Kedah, Pulau Pinang, Perak, Selangor, Kuala Lumpur, Negeri Sembilan, Melaka, Johor, Pahang, Terengganu, Kelantan	Zieritz et al., 2016
5.	Species list: *Sinanodonta woodiana, Pilsbryoconcha exilis, Pilsbryoconcha compressa, Pseudodon cambodjensis, Pseudodon cumingii, Pseudodon vondembuschianus, Contradens contradens, Hyriopsis bialata, Physunio superbus* and *Rectidens sumatrensis*.	Zieritz et al., 2016

Appendix 1 (Continued)

No.	Content	References
6.	34 species: *Anguilla borneensis, Anguilla marmorata, Barbodes sealei Barbonymus balleroides Crosseocheilus elegans, Cyclocheilichthys repasson, Garra borneensis, Hampala sabana, Leptobarbus melanotaenia, Lobocheilos erinaceus, Lobocheilos unicornis, Luciosoma pelligrinii, Nematobramis everetti, Osteochilus chini, Osteochilus ingeri, Rasbora cf. sumatrana, Rasbora elegans, Rasbora hubbsi, Tor tambra, Nemacheilus olivaceus, Hemibagrus baramensis, Hemibagrus fortis, Clarias infarcts, Gastromyzon ingeri, Gastromyzon lepidogaster, Parhomaloptera microstoma, Protomyzon borneensis, Protomyzon griswoldi, Betta ocellata, Channa striata, Macrognathus keithi, Mastacembelus unicolor, Homolopteroides stephensoni, Oreochromis mossambicus*	Wilkinson et al., 2018
7.	16 species: *Crossocheilus elegans, Garra borneensis, Lobocheilos erinaceus, Lobocheilos unicornis, Luciosoma pelligrinii, Betta ocellata, Mastacembelus unicolor, Homalopteroides stephensoni, Macrognathus keithi, Rasbora elegans; Protomyzon borneensis; Barbonymus balleroides, Clarias anfractus, Cyclocheilichthys repasson, and Channa striata and Oreochromis mossambicus*	Wilkinson et al., 2018
8.	List of sampling stations in Peninsular Malaysia: (I,ii) Mangrove area, Kuala Gula & Kuala Sepetang, Perak; (iii) River area of Bagan Lalang, Selangor; (iv) Irrigation canals area of Kuala Linggi, Negeri Sembilan; (v) Mangrove area of Tanjung Piai, Johor	Yusof et al., 2012

No.	Content	References
9.	31 species: 1 = *Ambassis interrupta;* 2 = *Nuchequula longicornis;* 3 = *Leiognathus equula;* 4 = *Siganus guttatus;* 5 = *Lutjanus russellii;* 6 = *Yarica hyalosoma;* 7 = *Dichotomyctere nigroviridis;* 8 = *Ambassis kopsii;* 9 = *Strongylura strongylura;* 10 = *Lates calcarifer;* 11 = *Pseudorhombus arsius;* 12 = *Epinephelus coioides;* 13 = *Chelonodontops patoca; 14 = Hexanematichthys sagor;* 15 = *Brachirus orientalis;* 16 = *Siganus javus;* 17 = *Pelates quadrilineatus;* 18 = *Arius maculatus;* 19 = *Zenarchopterus dunckeri;* 20 = *Glossogobius aureus;* 21 = *Vespicula trachinoides;* 22 = *Allenbatrachus grunniens;* 23 = *Psammogobius biocellatus;* 24 = *Sillago sihama;* 25 = *Toxotes jaculatrix;* 26 = *Triacanthus blochii;* 27 = *Gerres filamentosus;* 28 = *Pamadasys kaakan;* 29 = *Gerres limbatus;* 30 = *Platycephalus indicus;* 31 = *Brevitrygon walga.*	Kadir et al., 2020
10.	East coast Peninsular Malaysia: Pahang, Terengganu, and Kelantan	Yeny-Nadira et al., 2019
11.	20 species: *Anabas testudineus, Hemibagrus nemurus, Hemibagrus planiceps, Mystus nigriceps, Channa melasoma, Clarias leiacanthus, Barbonymus gonionotus, Barbonymus schwanefeldii, Cyclocheilichthys apogon, Cyclocheilichthys heteronema, Osteochilus vittatus, Parachela oxygastroides, Rasbora dusonensis, Rasbora tornieri, Helostoma temminckii, Notopterus notopterus, Trichopodus trichopterus, Trichopodus pectoralis, Pristolepis grootii, Wallago leerii*	Sule et al., 2018

Chapter 2

The Impacts of the Paleoenvironment on Fish Distribution

Min Pau Tan*
and Siti Azizah Mohd Nor
Institute of Climate Adaptation and Marine Biotechnology,
Universiti Malaysia Terengganu, Kuala Nerus, Terengganu, Malaysia

Abstract

The contemporary distribution of aquatic species is correlated with the historical population processes, life histories, and paleo- and current ecological arrangements associated with climatic changes. Specifically, the paleoclimatic changes during the Pleistocene Epoch in the Quaternary Period, which resulted in sea surface temperature and sea level oscillations, have been associated with major alterations in genetic diversity, population sub-division, distribution patterns, demographic history, and speciation events. Understanding the paleogeographical arrangement, aside from the current climatic conditions and anthropogenic factors, is thus important in elucidating the present-day genetic distribution among fish species. One of the key approaches used to comprehend the paleogeographical arrangement is molecular and bioinformatic tools. In this chapter, we review selected published articles and discuss the impacts of the paleoenvironment on shaping the distribution of fish taxa based on molecular data and bioinformatic analyses. This chapter highlights how the distribution of biodiversity was impacted unevenly by paleoclimatic changes over a large spatial scale. In particular, paleoecological factors played important roles in shaping the taxonomic distribution, life history, or population processes of

* Corresponding Author's Email: mptan@umt.edu.my.

In: Climate Change and Environmental Perturbations
Editors: T. Min Pau, M. Abd. Ghaffar, M. Danish-Daniel Abdullah et al.
ISBN: 979-8-89113-271-9

specific species, thus having a certain influence on their modern geographical ranges. The studies detailed here also illustrated the importance of covering a wider sampling range to provide a clearer understanding of the relationship between climate and paleoenvironmental changes and the evolution of biodiversity. Additionally, investigations involving multi-locus genetic markers provide a useful approach to supporting historical demographic analyses, helping to avoid the discordant results that may arise when relying on one or a few markers when reconstructing historical demographics. Similarly, a single-taxa investigation may underestimate a historical event; thus, the inclusion of multiple distinct taxa from a particular spatial range often yields more explicit conclusions. This chapter provides insight into the roles of the paleoenvironment and paleoclimatic changes in outlining the modern genetic distribution of diverse fish inhabiting various geographical regions.

Keywords: climate, genetic, biodiversity, aquatic

Introduction

The contemporary distribution of aquatic species is correlated with the historical population processes, life history, and paleo and current ecological arrangements associated with climatic changes (Ruzzante et al., 2011; Ali & Aitchison, 2014; Mohd Yusof et al., 2021). In particular, the paleoclimatic changes during the Pleistocene Epoch in the Quaternary Period, which resulted in sea surface temperature and sea level oscillations, have been linked to significant shifts in genetic diversity, population sub-division, distribution patterns, demographic history, and speciation events (Almada et al., 2017; Beck et al., 2017; de Carvalho Mondin et al., 2018; Sholihah et al., 2021). Sea levels fluctuations occurred during different geological periods in the past, leading to episodic marine invasions during deglaciation periods. In contrast, when the sea level dropped during glaciations, extended shorelines and land bridges were formed. The unfavourable climate and frequent rising and falling of sea levels led to land mass reformations, further shaping the movement and expansion/retraction of populations into isolated refugia (Shafer et al., 2010; Tan et al., 2012; 2015; Campbell et al., 2015).

During the Last Glacial Maximum, a major landscape alteration was observed in many parts of the continent. For instance, within the megadiverse Southeast Asia region, the Sunda shelf was once an exposed lowland that bridged Java, Malay Peninsula, Sumatra. Borneo, Madura, Bali, and surrounding smaller islands into a major landmass. However, the increased sea water levels as a consequence of deglaciations in the late glacial period inundated the lowland areas of the Sunda shelf, leaving disjointed land masses (Voris, 2000). At present, the Malay Peninsula is separated from the island of Sumatra (Indonesia) by the Straits of Malacca and disconnected from the island of Borneo by the South China Sea. Such ecological rearrangements have affected the distribution and recolonization of the freshwater ichthyofauna, as these obligate freshwater taxa are restricted from dispersion by seawater (Tan et al., 2012; 2015, Beck et al., 2017; Jamaluddin et al., 2019). Individuals surviving in fragmented habitats might have evolved independently and, thus, have different evolutionary paths from each other. Consequently, the study of spatial genetic patterning among these isolated survivors could shed light on the past ecological processes and demographic history.

In the Northeastern Atlantic, intensive climatic changes during the Pleistocene epoch significantly influenced the spatial distributions of marine communities (Robalo et al., 2011; Almada et al., 2017). When the glaciation cycles of the North Sea occurred, resulting in a reduced sea water temperature, marine populations retreated into southern regions or survived in northern peri-glacial refugia (Maggs et al., 2008; Robalo et al., 2011; Briggs & Bowen, 2012). Previous studies have been conducted to identify the interglacial/postglacial refugia where temperate fish survived the successive glacial peaks, in order to comprehend the mechanism of recolonization and population expansion, when conditions permit (Wares & Cunningham, 2001; Jenkins et al., 2018).

Understanding the paleogeographical arrangement, aside from the current climatic condition and anthropogenic factors, is important in elucidating the present-day genetic distribution among fish species. One of the key approaches used to comprehend the paleogeographical arrangement is through the use of molecular and bioinformatic tools. Bioinformatic tools are computer software programs developed for the extraction of meaningful information from biological databases. Specifically, advancements in molecular approaches and bioinformatics tools have provided opportunities to understand the ecological, evolutionary, and historical mechanisms that have influenced genetic patterns. Multiple molecular genetic studies using different nuclear and mitochondrial DNA target regions have been documented for many aquatic species inhabiting various habitats. In this chapter, we review selected published articles and discuss the impacts of the paleoenvironment in shaping the distribution of fish taxa, based on molecular data and bioinformatic analyses. This chapter provides insight into the roles of the paleoenvironment and paleoclimatic changes in outlining the modern genetic distribution of diverse fish inhabiting various geographical regions,

Paleoenvironmental Impacts on Freshwater Fishes

The complex climatic and geological history of Southeast Asia has shaped the biodiversity distribution within this region (Beck et al., 2017). In particular, the rising and lowering sea levels associated with recurrent glacial cycles during the Pleistocene Epoch have both eased and restricted the integration of populations. For example, in a study on a widespread fresh/brackish water killifish (*Aplocheilus panchax*), three major mitochondrial lineages were revealed, which were associated to divergence events occurring during the Pliocene and Pleistocene periods (Beck et al., 2017). The genetic variation of fish populations signals a West–East migration, followed by rapid expansion across Southeast Asia. Individuals from Krabi (north of Isthmus of Kra, Thailand) were genetically closer to populations from the Indian Ocean, suggesting the delineation of a freshwater taxa impediment at the Isthmus of Kra biogeographic barrier. This study also demonstrated a retention of historical isolation within the maternal lineages of *A. panchax* populations, in addition to their capacity to disperse across considerable geophysical barriers in relative recent time periods, as indicated by nuclear genetic variation.

A signature of ancient river connectivity within Sundaland was evidenced through a phylogeographic study of the striped snakehead (*Channa striata*) (Tan et al., 2012; 2015). In particular, the genetic closeness between samples from the western Sarawak state of Malaysian Borneo, southern Peninsular Malaysia, east Peninsular Malaysia, and Sumatra (Indonesia) based on mitochondrial DNA markers provided evidence of a paleo North Sunda River system that existed when sea level was way below that at present (Figure 1). In addition, the sharing of haplotype between fish from Sumatra and east Peninsular Malaysia suggested a historical connection between these regions by the ancient Siam River systems, which enabled long-distance movement across these regions. Similarly, genetic proximity between these two regions was also observed in climbing perch (*Anabas testudineus*), which naturally traversed the ancient river system during the most recent Pleistocene glaciations (Jamsari et al., 2010).

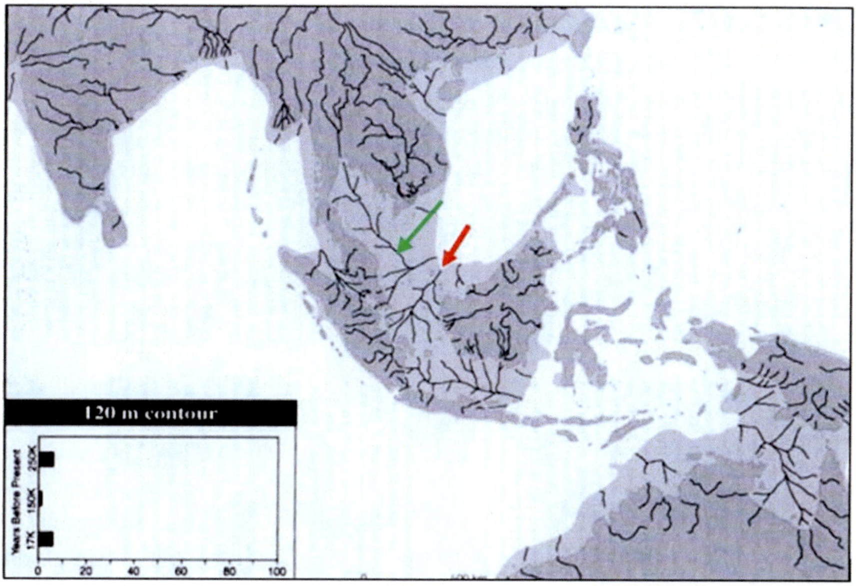

Figure 1. The ancient North Sunda River system (red arrow) and Siam River system (green arrow) (modified from Voris (2000), Figure 1(a), p. 1156).

On the other side of the continent, the Neotropics (or Neotropical realm) is among one of the terrestrial realms that contains the highest amount of biodiversity on earth. The fish diversity distribution within this region is highly heterogenous, due to landscape rearrangements resulting from complex geologic and climatic histories, such as the plate tectonic collisions (Albert et al., 2006) and the uplift of the Andes (Salgado-Roa et al., 2018). Together, these historical processes may have affected dispersal patterns, influenced geographical fragmentation, and ultimately promoted speciation (Turchetto-Zolet et al., 2013; Hazzi et al., 2018; Salgado-Roa et al., 2018). A recent study by Rincon-Sandoval et al. (2019), which aimed at revealing the co-diversification patterns of freshwater fishes through the use of a genomic comparative phylogeographic method, provided new insights into historical ecological reconfigurations associated with major shifts in riverine systems, resulting in co-diversification and speciation over geologic time.

Climate changes during the Pleistocene also significantly affected the genetic diversity, composition, distribution, and population size of Arctic organisms (Hewitt, 2000; Campbell et al., 2015). In a study on the blackfish *Dallia pectoralis* and *D. admirabilis*, which are restricted entirely to Beringia, a total of five distinct genetic clusters corresponding to the aquatic sub-refugia in Beringia were evident, due to the strictly freshwater nature and low dispersal ability of these species (Campbell et al., 2015). It appears that the distribution range of *Dallia* species was restricted by paleoclimatic variation, where the *Dallia* were once more widespread prior to the Pleistocene glaciations, and they did not re-expand at the end of the glaciation. It is worth noting that a post-Pleistocene range expansion is classic for many fish species. Nevertheless, changes in climate have harshly affected the survival and spatial dispersion of *Dallia* in many parts of its range, leading to population fragmentation and, subsequently, population and genetic bottlenecks. Migration rates between the sub-refugia were at low levels, further isolating individuals from different geographical areas. In the same study, Paleoclimatic instability was found to facilitate intercontinental exchange (i.e., between Asia and North America), as opposed to generating species divisions between continents.

Although the genetic patterning of freshwater species was attributed to the barriers imposed by riverine environments, studies have shown that the co-distributed taxa of freshwater fish with strong spatial concordance might display different genetic patterns, given the species-specific ecological associations and dispersal capabilities (Jenkins et al., 2018). Particularly, different species or taxa capitalize differently on interim dispersal pathways that became available during Pleistocene sea-level invasions and retreats (Thomas & Knowles, 2020). This was evident in the coastal riverine characid fish community (Ostariophysi: Characiformes), commonly known as tetras, of the Brazilian Atlantic Rainforest. Genomic tests on the four co-distributed characid taxa—*Mimagoniates microlepis*, *Hyphessobrycon boulengeri, Hollandichthys multifascaitus,* and *Bryconamericus microcephalus*—indicated a common route for dispersal; however, each of the fish species dispersed at different temporal intervals, resulting in different genetic distribution patterns across barriers, with most divergences occurring during the Upper Pleistocene (Thomas & Knowles, 2020).

In a separate study, Jenkins et al. (2018), in their meta-analysis using the mitochondrial DNA data set of a group of co-distributed taxa, revealed contrasting phylogeographic patterns despite the commonality of post-Last Glacial Maximum expansion patterns in many of the studied taxa. These findings implicate the necessity to develop specific predictions at the taxon level. Thus, understanding the adaptive response of each organism to these extrinsic and intrinsic factors, and how they influence population segregation, is crucial for the generation of estimates regarding their response to future sea level increases.

Paleoenvironmental Impacts on Marine Fishes

In the marine realm, the contemporary genetic distribution and connectivity of a species are driven by both past and present barriers to gene flow. Oceanographic features, such as currents, fronts, and upwelling systems, can be associated to the dispersal of a marine species at one or all of its life stages, further shaping the population sub-structure. This has been observed in a migratory and benthopelagic sciaenid fish (*Atractoscion aequidens*) inhabiting Benguela in the Southeastern Atlantic (Henrique et al., 2014). Using the mitochondrial DNA D-loop and short sequence repeat markers, populations of *A. aequidens* were found to have historical and significant isolation and divergence, which is apparent up to the present time. These results indicate that the Benguela Current system may have limited the genetic connectivity for warm-temperate fish taxa. However, further studies on other species in similar spatial locations are needed to support this hypothesis.

Marine biogeographic barriers can have contrasting consequences, even among sister species that share similar traits in morphology, life history, ecology, and distribution ranges (DiBattista et al. 2012). This is due to the complex historical population processes related to the extinction–recolonization processes that may have potentially masked phylogeographic signals correlated with biogeographic barriers in the marine environment. Specifically, one species can be more vulnerable to population contraction and extinction events over time. An example is the Indo-Pacific reef butterflyfish (*Chaetodon meyeri* and *C. ornatissimus*). Despite the high resemblance in life history traits and capability of interbreeding among them (Hobbs et al., 2009), they responded differently in shared environmental, ecological, and evolutionary regimes. Populations of both species expanded during the Pleistocene, yet *C. meyeri* appears to have expanded much more recently than *C. ornatissimus*, reinforcing that their evolutionary history may differ, as has also been reported for other Chaetodontidae species (Lawton et al., 2011). From the same study, the mitochondrial and microsatellite markers did not provide concordant results for *C. meyeri*, or in 7 out of 12 other cases of marine fishes in the published literature (Lawton et al., 2011). The contrasting genetic signals serve as a reminder that caution should be exercised when relying on a single or a few markers to reconstruct historical demographics.

Our ability to identify biodiversity hotspots and to understand the mechanisms that produce them are critical for already-challenging conservation efforts. In a study of an economically important grouper species (*Cephalopholis argus*), genetic analyses showed that this species was significantly affected by the episodic Pleistocene sea level changes, which sub-divided this species into the Pacific and Indian Oceans (Gaither et al, 2011). The significant genetic structuring was contributed to by the profound Indo–Pacific Barrier (IPB), especially during the low sea level associated with glaciation events. The IPB also played a significant role in hindering gene flow between populations of the two-spined yellowtail stargazer (*Uranoscopus cognatus*) from the west of Thailand and east coast of Peninsular Malaysia, thus partitioning them into two separate management units (Mohd Yusoff et al., 2021).

Historical events during the Pleistocene Epoch may have influenced the genetic diversification of longtail tuna (*Thunnus tonggol*) populations from the Indo-West Pacific region (Syahida Kasim et al., 2020). During the interglacial and/or deglaciation periods, the increased sea levels provided larger and more suitable habitats for this pelagic and migratory species. Results from all analyses, such as the neutrality tests, sum of squared deviations, Harpending's raggedness index, median-joining network, mismatch distribution, and estimation of population size before and after expansion, all supported a historical population expansion event in longtail tuna populations, dated to the Pleistocene era (Syahida Kasim et al., 2020).

In another study, conducted to understand the evolution of a widespread amphidromous Indo-Pacific goby (*Sicyopterus lagocephalus*) using several coalescence-based methods, three major haplogroups were found, where each was specific to a biogeographic region (Hoareau et al., 2012). These populations indicated the impacts of demographic contraction–expansion events associated to Pleistocene sea-level changes. The heterogenous environmental forces resulting from climatic changes have impacted marine organisms in two ways: first, the lowering of sea levels during glacial periods isolated the marine habitats, preventing or reducing the chance for secondary contact and, thus, resulting in genetic differentiations. Second, when sea levels increased during the melting of ice sheets at interglacial periods, range and population expansions were facilitated as populations gained larger and more suitable habitat areas (Provan and Bennett, 2008; Marko et al., 2010; Hoareau et al., 2012).

Conclusion

In conclusion, the distribution of biodiversity was impacted unevenly by paleoclimatic changes over a large spatial scale. Though paleoecological factors appear to have played important roles in shaping the taxa distribution, life history, or population processes of specific species, thus having a certain influence on their modern geographical ranges, the studies detailed here also illustrated the importance of covering a wider sampling range to provide a clearer understanding of the relationship between climate and paleoenvironmental changes and the evolution of biodiversity. Additionally, investigations involving multi-locus genetic markers provide a useful approach to support historical demographic analyses, helping to avoid the discordant results which may arise when relying on one or a few markers when reconstructing historical demographics. Similarly, a single-taxa investigation may underestimate a historical event; thus, the inclusion of multiple distinct taxa from a particular spatial range often yields more explicit conclusions. Bioinformatic tools are powerful in elucidating the paleoenvironmental impacts on the distribution of fish taxa. Incorporating other data, such as marine isotope stages (MIS) for marine fish studies and fossil records, will provide further insight into the relationship between paleoenvironmental rearrangements and biodiversity distribution.

References

Albert, J. S., Lovejoy, N. R., and Crampton, W. G. R. (2006). Miocene tectonism and the separation of cis- and trans-Andean River basins: Evidence from Neotropical fishes. *Journal of South American Earth Sciences*, 21(1–2), 14–27. https://doi.org/10.1016/j.jsames.2005.07.010.

Ali, J. R., and Aitchison, J. C. (2014). Exploring the combined role of eustasy and oceanic island thermal subsidence in shaping biodiversity on the Galápagos. *Journal of Biogeography*, 41(7), 1227-1241.

Almada, F., Francisco, S. M., Lima, C. S., FitzGerald, R., Mirimin, L., Villegas-Rios, D., Saborido-Rey, F., Afonso, P., Morato, T., Bexiga, S., and Robalo, J. I. (2017). Historical gene flow constraints in a northeastern Atlantic fish: phylogeography of the ballan wrasse *Labrus bergylta* across its distribution range. *Royal Society Open Science*, 4, 160773. http://doi.org/10.1098/rsos.160773.

Beck, S. V., Carvalho, G. R., Barlow, A., Rüber, L., Tan H. H., Nugroho E, Wowor, D., Mohd Nor, S. A., Herder, F., Muchlisin, Z. A., and De Bruyn, M. (2017). Plio-Pleistocene phylogeography of the Southeast Asian Blue Panchax killifish, *Aplocheilus panchax*. *pLoS ONE*, 12(7), e0179557. https://doi.org/10.1371/journal.pone.0179557.

Briggs, J. C., and Bowen, B. W. (2012). A realignment of marine biogeographic provinces with particular reference to fish distributions. *Journal of Biogeography*, 39, 12–30. http://dx.doi.org/10.1111/j.1365-2699.2011.02613.x.

Campbell, M. A., Takebayashi, N., and López, J. A. (2015). Beringian sub-refugia revealed in blackfish (*Dallia*): implications for understanding the effects of Pleistocene glaciations on Beringian taxa and other Arctic aquatic fauna. *BMC Evolutionary Biology*, 15, 144. https://doi.org/10.1186/s12862-015-0413-2.

de Carvalho Mondin, L.A., Machado, C. B., de Resende, E. K., Marques, D. K. S., and Galetti Jr, P. M. (2018). Genetic pattern and demographic history of *Salminus brasiliensis*: Population expansion in the Pantanal region during Pleistocene. *Frontiers in Genetics*, 8. https://doi.org/10.3389/fgene.2018.00001.

DiBattista, J. D., Rocha, L. A., Craig, M. T., Feldheim, K. A., and Bowen, B. W. (2012). Phylogeography of two closely related Indo-Pacific butterflyfishes reveals divergent evolutionary histories and discordant results from mtDNA and microsatellites, *Journal of Heredity*, 103(5), 617–629, https://doi.org/10.1093/jhered/ess056.

Field, R., Hawkins, B. A., Cornell, H. V., Currie, D. J., Diniz-Filho, J. A. F., Guégan, J. F. Kaufman, D. M., Kerr, J. T., Mittelbach, G. G., Oberdorff, T., and O'Brien, E. M. (2009). Spatial species-richness gradients across scales: a meta-analysis. *Journal of Biogeography*, 36, 132–147.

Gaither, M. R., Bowen, B. W., Bordenave, T. R., Rocha, L. A., Newman, S. J., Gomez, J. A., van Herwerden, L., and Craig, M. T. (2011). Phylogeography of the reef fish *Cephalopholis argus* (Epinephelidae) indicates Pleistocene isolation across the indo-pacific barrier with contemporary overlap in the coral triangle. *BMC Evolutionary Biology*, 11, 189. https://doi.org/10.1186/1471-2148-11-189.

Hazzi, N. A., Sebastián, J., Ortiz-movliav, C., and Darío, R. (2018). Biogeographic regions and events of isolation and diversification of the endemic biota of the tropical Andes. *Proceedings of the National Academy of Sciences of the United States of America*, 115(31), 7985– 7990. https://doi.org/10.1073/pnas.1803908115.

Henrique, R., Potts, W. M., Santos, C. V., Sauer, W. H. H., and Shaw, P. W. (2014). Population connectivity and phylogeography of a coastal fish, *Atractoscion aequidens* (Sciaenidae), across the Benguela Current Region: evidence of an ancient vicariant event. *pLoS ONE*, 9(2), e87907. https://doi.org/10.1371/journal.pone.0087907.

Hewitt, G. (2000). The genetic legacy of the Quaternary ice ages. *Nature*, 405, 907–13.

Hoareau, T. B., Boissin, E., and Berrebi, P. (2012). Evolutionary history of a widespread Indo-Pacific goby: The role of Pleistocene sea-level changes on demographic contraction/expansion dynamics. *Molecular Phylogenetics and Evolution*, 62(1), 566-572. https://doi.org/10.1016/j.ympev.2011.10.004.

Hobbs, J. P., Frisch, A. J., Allen, G. R., and Van Herwerden, L. (2009). Marine hybrid hotspot at Indo-Pacific biogeographic border. *Biology Letters*, 5, 258–261.

Jamaluddin, J. A. F., So, N., Tam, B. M., Ahmad, A., Grudpan, C., Page, L. M., Khaironizam, M. Z., and Mohd Nor, S. A. (2019). Genetic variation, demographic history and phylogeography of tire track eel, *Mastacembelus favus* (Synbranchiformes: Mastacembelidae) in Southeast Asia. *Hydrobiologia*, 838, 163–182.

Jamsari, A. F. J., Muchlisin, Z. A., Musri, M., and Siti Azizah, M. N. (2010). Remarkably low genetic variation but high population differentiation in the climbing perch, *Anabas testudineus* (Anabantidae), based on mtDNA control region. *Genetics and Molecular Research*, 9(3), 1836–1843.

Jenkins, T. L., Castilho, R., and Stevens, J. R. (2018). Meta-analysis of northeast Atlantic marine taxa shows contrasting phylogeographic patterns following post-LGM expansions. *PeerJ*, 6, e5684 https://doi.org/10.7717/peerj.5684.

Lawton, R. J., Messmer, V., Pratchett, M. S., and Bay, L. K. (2011). High gene flow across large geographic scales reduces extinction risk for a highly specialised coral feeding butterflyfish. *Molecular Ecology*, 20, 3584–3598.

Maggs, C. A., Castilho, R., Foltz, D., Henzier, C., Jolly, M. T., Kelly, J., Olsen, J., Perez, K. E., Stam, W., Väinölä, R., and Viard, F. (2008). Evaluating signatures of glacial *refugia* for North Atlantic benthic marine taxa. *Ecology*, 89, S108–S122.

Marko, P. B., Hoffman, J. M., Emme, S. A., McGovern, T. M., Keever, C. C., and Cox, L. N. (2010). The 'Expansion-Contraction' model of Pleistocene biogeography: rocky shores suffer a sea change? *Molecular Ecology*, 19, 146-169.

Mohd Yusoff, N. I. S., Mat Jaafar, T. N. A., Vilasri, V., Mohd Nor, S. A., Seah, Y. G., Habib, A., Wong, L. L., Danish-Daniel, M., Sung, Y. Y., Mazlan, A. G., and Mat Piah, R. (2021). Genetic diversity, population structure and historical demography of the two-spined yellowtail stargazer (*Uranoscopus cognatus*). *Scientific Reports*, 11, 13357. https://doi.org/10.1038/s41598-021-92905-6.

Provan, J., and Bennett, K. D. (2008). Phylogeographic insights into cryptic glacial refugia. *Trends in Ecology & Evolution*, 23, 564-571.

Rincon-Sandoval, M., Betancur-R, R., and Maldonado-Ocampo, J. A. (2019). Comparative phylogeography of trans-Andean freshwater fishes based on genome-wide nuclear and mitochondrial markers. *Molecular Ecology*, 28(5), 1096-1115.

Robalo, J. I., Castilho, R., Francisco, S. M., Almada, F., Knutsen, H., Jorde, P. E., Pereira, A. M., and Almada, V. C. (2011). Northern refugia and recent expansion in the North Sea: the case of the wrasse *Symphodus melops* (Linnaeus, 1758). *Ecology and Evolution*, 2(1), 153-164. https://doi.org/10.1002/ece3.77.

Ruzzante, D. E., Walde, S. J., Macchi, P. J., Alonso, M., and Barriga, J. P. (2011). Phylogeography and phenotypic diversification in the Patagonian fish *Percichthys trucha*: the roles of Quaternary glacial cycles and natural selection, *Biological Journal of the Linnean Society*, 103(2), 514–529. https://doi.org/10.1111/j.1095-8312.2011.01682.x.

Salgado-Roa, F. C., Pardo-Diaz, C., Lasso, E., Arias, C. F., Solferini, V., and Salazar, C. (2018). Gene flow and Andean uplift shape the diversification of Gasteracantha cancriformis (Araneae: Araneidae) in Northern South America. *Ecology and Evolution*, 8(14), 7131– 7142. https://doi.org/10.1101/250977.

Shafer, A. B. A., Cullinham, C. I., Côté, S. D., and Coltman, D. W. (2010). Of glaciers and refugia: a decade of study sheds new light on the phylogeography of northwestern North America. *Molecular Ecology*, 19, 4589–621.

Sholihah, A., Delrieu-Trottin, E., Condamine, F. L., Wowor, D., Rüber, L., Pouyaud, L., Agnèse, J. F., and Hubert, N. (2021). Impact of Pleistocene Eustatic Fluctuations on Evolutionary Dynamics in Southeast Asian Biodiversity Hotspots. *Systematic Biology*, 70(5), 940–960, https://doi.org/10.1093/sysbio/syab006.

Syahida Kasim, N., Mat Jaafar, T., Mat Piah, R., Mohd Arshaad, W., Mohd Nor, S. A., Habib, A., Abd Ghaffar, M., Sung, Y. Y., Danish-Daniel, M., and Tan, M. P. (2020). Recent population expansion of longtail tuna *Thunnus tonggol* (Bleeker, 1851) inferred from the mitochondrial DNA markers. *Peer J*, 8, e9679. https://doi.org/10.7717/peerj.9679.

Tan, M. P., Jamsari, A. F. J., and Siti Azizah, M. N. (2012). Phylogeographic pattern of the striped snakehead, *Channa striata* in Sundaland: Ancient river connectivity, geographical and anthropogenic signatures. *pLoS ONE*, 7(12), e52089. https://doi.org/10.1371/journal.pone.0052089.

Tan, M. P., Jamsari, A. F. J., Muchlisin, Z. A., and Siti Azizah, M. N. (2015). Mitochondrial genetic variation and population structure of the striped snakehead, *Channa striata* in Malaysia and Sumatra, Indonesia. *Biochemical Systematics and Ecology*, 60, 99-105. https://doi.org/10.1016/j.bse.2015.04.006.

Thomas, A. T., and Knowles, L. L. (2020). Common barriers, but temporal dissonance: Genomic tests suggest ecological and paleo-landscape sieves structure a coastal riverine fish community. *Molecular Ecology*, 29(4), 783-796.

Turchetto-Zolet, A. C., Pinheiro, F., Salgueiro, F., and Palma-Silva, C. (2013). Phylogeographical patterns shed light on evolutionary process in South America. *Molecular Ecology*, 22(5), 1193–1213. https://doi.org/10.1111/mec.12164.

Voris, H. K. (2000). Maps of Pleistocene sea levels in Southeast Asia: shorelines, river systems and time durations. *Journal of Biogeography*, 27(5), 1153-1167. https://doi.org/10.1046/j.1365-2699.2000.00489.x.

Wares, J. P., and Cunningham, C. W. (2001). Phylogeography and historical ecology of the North Atlantic intertidal. *Evolution*, 55, 2455–2469.

Chapter 3

Culture-Dependent and Independent Approaches to Studying the Impact of Bacterial Communities Associated with Harmful Algal Blooms

Muhd Danish-Daniel[1,*]
Mohd Ezhar Mohd Noor[2]
Min Pau Tan[1]
and Yeong Yik Sung[1]
[1]Institute of Climate Adaptation and Marine Biotechnology, Universiti Malaysia Terengganu, Kuala Nerus, Terengganu, Malaysia
[2]Department of Agrotechnology and Bio-Industry, Jeli Polytechnic, Jeli, Kelantan, Malaysia

Abstract

Harmful algal blooms (HABs) are recurring phenomena that pose serious threat to human health and aquatic ecosystems. Bacterial populations connected to HABs are fundamental in determining the dynamics and ecological effects of these phenomena. Understanding the composition, diversity, and functional roles of these bacterial communities is crucial for the development of effective monitoring and management strategies. This chapter provides an overview of culture-dependent and culture-independent approaches used to study the bacterial community associated with HABs. Traditional cultivation methods are used to isolate and characterise microorganisms from environmental samples. With the

* Corresponding Author's Email: mdda@umt.edu.my.

In: Climate Change and Environmental Perturbations
Editors: T. Min Pau, M. Abd. Ghaffar, M. Danish-Daniel Abdullah et al.
ISBN: 979-8-89113-271-9

help of these techniques, particular bacterial strains that may be directly responsible for HAB development or mitigation can be identified and characterised. Meanwhile, culture-independent approaches have become more popular recently. These methods include molecular-based techniques such as next-generation sequencing, metagenomics, and metatranscriptomics. These techniques offer a comprehensive and unbiased assessment of the bacterial population linked to HABs by directly analysing the genetic material present in environmental samples. They make it possible to identify uncommon or difficult-to-culture bacteria and provide insights into the community's functional potential and metabolic capabilities. This chapter also discusses the advantages and limitations of both culture-dependent and culture-independent approaches to unveil the interaction complexity of bacterial communities associated with HABs.

Keywords: HABs, microbial community, bacteria-algal interactions, metagenomics

Introduction

Microalgae are the major producers of biomass and organic compounds in the ocean. More than 5000 species of marine microalgae are known at present, separated into six major divisions: *Chlorophyta* (green algae), *Ochrophyta* (yellow algae, golden brown, and diatoms), *Haptophyta* (coccolithophorids), *Pyrrhophyta* (dinoflagellates), *Euglenophyta,* and *Cyanophyta* (blue–green algae) (Hallegraef, 1995). Among these 5000 species, about 300 can proliferate in high numbers to form the so-called red tide and brown tide phenomena (Daranas et al., 2001; Hallegraeff, 1995).

Many planktonic organisms can form mass occurrences in the water column. When the cell densities reach values considerably higher than their general background distribution, they are called blooms (Smayda, 1997). Blooms can be almost mono-specific, while others are formed by a combination of species (Popovich et al., 2008). Many prominent blooms can be traced back to high nutrient loads (Yentsch et al., 2008), but they occur whenever a species is able to outgrow its competitors while partially reducing grazer pressure (Irigoien et al., 2005).

As the base of marine food chains, many phytoplankton blooms are beneficial for ocean productivity, hence benefiting the fishing and shellfish industries (Hallegraeff, 1993) (Figure 1). Other blooms, however, often consist of dinoflagellates, which can significantly disrupt ecosystems (Wear and Gardner, 2001), harm commercially important species (Tang and Gobler, 2009), or produce substances which are toxic to humans (Van Dolah, 2000). Coastal eutrophication may, in some cases, lead to an increase in such harmful bloom events. While part of the apparent global increase in harmful algal blooms (HABs) might be due to increased scientific awareness and increased commercial activities in coastal areas, eutrophication, climatic effects, and the introduction of new harmful species (e.g., by ship ballast water or through the import of shellfish stocks for aquaculture) are likely to alter the current patterns of bloom occurrence (Leung and Dudgeon, 2008).

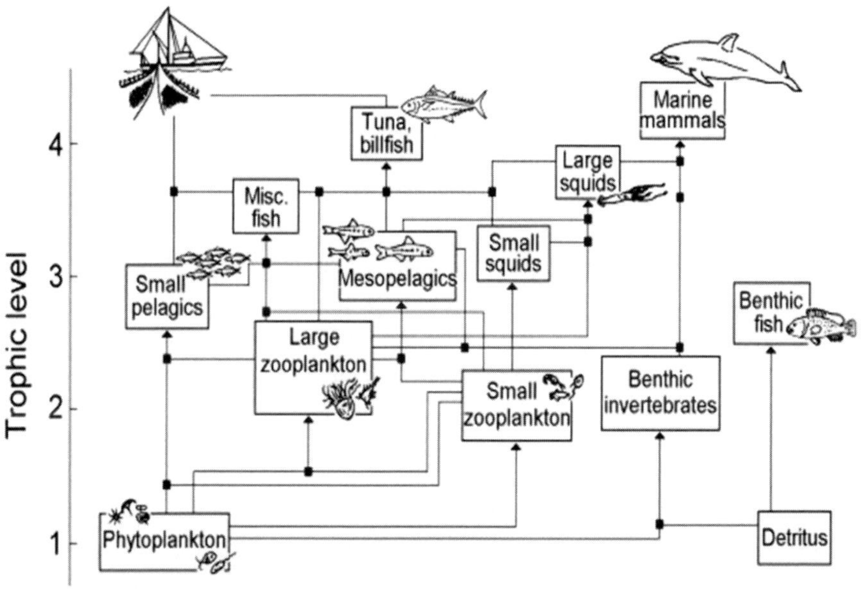

Figure 1. A simplified food web defining the various elements of such webs ('functional groups'), the flows between them, and so-called 'trophic levels,' indicating the position of each functional group within the web which benefits the fishery and shellfish industries (Pauly, 2003; Hallegraeff, 1993).

Dinoflagellates are a great diverse group of protists that can be found in most aquatic habitats worldwide. Presenting a wide diversity of lifestyles, they are major components of both phytoplankton and micro-grazer communities. Photosynthetic dinoflagellates are among the most important phytoplankton groups in marine ecosystems. While exact data on their global biomass production are not available, they occur from the Arctic and Antarctic marine ice brine channels (Thomson et al., 2006) to the tropics (Parab et al., 2006). In some instances, dinoflagellates can regularly or even permanently (Landry and Kirchman, 2002) dominate the eukaryotic phytoplankton in the marine environment.

Cultures of dinoflagellates contain a considerable amount of bacteria, which likely accompanied the dinoflagellates in the original sample. Interactions between algae and bacteria are commonly observed in both freshwater and marine ecosystems, with bacteria increasingly cited as potentially important regulators in the initiation, maintenance, and decline process of algal blooms (Doucette and Tricks, 1995). This community of bacteria is believed to be associated with each dinoflagellate, which is also believed to contribute to the physiology and toxigenesis of the dinoflagellates (Doucette and Tricks, 1995).

Interactions between bacteria and phytoplankton such as dinoflagellates play an important role in regulating dinoflagellate toxin production. Previous studies on the interactions between bacteria and dinoflagellates have shown them to be highly variable and only sometimes specific. The effects of bacteria on toxic dinoflagellates include negative effects, such as cell lysis and death (Lovejoy et al., 1998), and positive effects, such as growth enhancement following the addition of bacteria to cultures (Ferrier et al., 2002). Examples of factors which may cause stimulation or inhibition by bacteria include the production of co-factors and secretion of signalling molecules controlling cellular processes in dinoflagellates (Gallacher and Smith 1999). In addition, bacterial influences on nutrient availability may result in the stimulation or inhibition of the growth or toxin production of dinoflagellates. Both toxin production by dinoflagellates and bacteria associated with toxic or non-toxic dinoflagellates have been documented; for example, Gallacher et al., (1997) provided evidence of paralytic shellfish toxin (PST) production by bacteria associated with dinoflagellate cultures.

Microbial Community Associated with Algae

Microscopic marine algae can be vectors of microbial communities, as they are universally associated in the ocean. In nature, most microbial communities are found to be adhered to microalgae, organisms, and inanimate surfaces. These interactions are dynamic and are important factors in microbial proliferation and survival. Aquatic algae in situ, as well as in laboratory culture conditions, are often found to be associated with a variety of bacterial strains

(Berland et al., 1970). Bacterial communities can be defined as multi-species bacteria assemblages in which organisms live together in a contiguous environment (host) and interact with each other (Konopka, 2009). Bacteria reproduce asexually, are sized between 0.1 to 20 μm, and can be rod-, cocci-, or comma-shaped.

Bacteria are universally associated with algae in the ocean. It has been reported that *Aeromonas*, *Alteromonas*, *Bacillus*, *Flavobacterium*, *Moraxella*, *Pseudomonas*, *Roseobacter,* and *Vibrio* are the bacterial genera most frequently associated with toxic dinoflagellates (Ashton, 2003). Some of these bacteria also produce tetrodotoxins (Ashton, 2003). The toxins present in the dinoflagellate might be the result of variable microbial symbionts, environmental conditions, or a combination of both influences (Tosteson et al., 1998). Thus, the interactions between bacteria and macroalgae are dynamic and important factors affecting microbial proliferation and survival (Tosteson et al., 1988); for example, Tosteson et al., (1989) found that the toxic dinoflagellate *Ostreopsis lenticularis* hosts a variety of symbiotic bacterial flora, but each dinoflagellate clonal culture sequesters a variable set of bacterial strains.

Associated bacterial strains have been reported to be involved in the toxicity development of cultured *O. lenticularis* (González et al., 1995). It has been determined that culturable bacteria from the genus *Pseudomonas*/*Alteromonas* were required for the production of enhanced toxicity in clonal laboratory cultures of *Ostreopsis* cells during the static phase of growth (González et al., 1995). They also reported that the number and diversity of associated bacterial genera was variable among dinoflagellate clones.

Approaches to Study Microbial Community in the Environment

Several methods have been utilized to study the bacterial communities associated with marine organisms. Culture-dependent methods are procedures used for the identification of bacteria. These methods have a limited capacity to characterize bacteria (Ivone Vaz-Moreira, 2011), as the cultivation of bacteria requires several types of agar and not all bacteria are culturable. Alternatively, culture-independent methods can be applied to circumvent the need for culturing. Culture-independent methods are more diverse and allow for more complex interactions with the host, compared with culture-dependent methods (Ryota Ishino, 2012). Bacterial community studies are relying more and more on culture-independent methods based on the direct isolation of bacterial DNA or RNA from their host without any culturing process, making it possible to study the total community from marine organism samples (Barbier, 2008).

Culture-Dependent Approach

Culture-dependent and culture-independent approaches are used to study prokaryotic populations existing in environmental samples. The culture-dependent approach employs artificial media in order to culture or enrich groups of micro-organisms inhabiting environmental samples (Horner-Devine, 2003). This approach is also used for the isolation of single colonies. Until recently, this method was considered standard and was carried out in most microbial ecology studies. (Horner-Devine, 2003). In culture-dependent methods, viable culturable cells can be recovered through the plate count method. After incubation, the colonies that form are assumed to result from a single cell; thus, the total CFUs per volume of sample can be calculated. The most probable number (MPN) technique is similar to the plate count method, except that this method depends on the detection of certain qualitative characteristics of the organism of interest; in particular, the growth of the target organism(s) in a liquid medium (Horner-Devine, 2003). Micro-organisms capable of being cultured are preserved in laboratory culture and strain collections. These organisms are classified by phenotypic and genotypic characteristics, and are ultimately assigned to a specific taxonomic group. Although culture-dependent techniques facilitate an excellent understanding of the physiological potential of the isolated organisms, they do not necessarily offer insight into the complete makeup of a microbial community (Horner-Devine, 2003).

Culture-dependent methods are biased, in the way that a micro-organism can only be cultivated after its physiological niche is evident and experimentally reproduced (Staley and Konopka, 1985). Moreover, it is widely accepted that culture-dependent studies can only assess between 0.001% and 15% of the actual microbial communities present in a sample (Amann et al., 1995). Most of the viable cells that can be viewed microscopically but are not considered culturable can be classified into two groups: the first group is species that have entered an unculturable state, or for which the cultivation conditions utilized are not adequate; while the second group consists of unknown species that, to date, have not been able to be cultured due to a lack of appropriate processes (Amann et al., 1995).

The culture-dependent method is relevant when applied to soil microbiology, as it offers an overall fraction of the microbes residing in specific soil. Although the research executed in this study focuses mainly on culture-dependent techniques - and, in turn, is influenced by the biases mentioned above - it is possible that, in environments having extreme conditions, such as the Atacama Desert, culturable organisms indicate more of the community than in less extreme environments. Culture-dependent methods provide invaluable knowledge of any previously undescribed environments and, with regard to such studies, it is necessary to perform culture-dependent studies, as this provides a solid backbone for future research applying culture-independent protocols.

Culture-Independent Approach

There are many culture-independent methods that are presently available, including DGGE, 16S rDNA clone library and, most recently, 16S metagenomic analysis. For the present study, 16S rDNA clone library and 16S metagenomic analysis were conducted in order to reveal the bacterial diversity and community structure associated with dinoflagellates. In recent years, it has been discovered that the majority of micro-organisms in the environment are not culturable by traditional plate culture methods. Considering this fact, culture-independent approaches are essential for the evaluation of microbial diversity in various types of environments (Amann et al., 1995). Through culture-independent approaches, scientists have been able to identify unculturable prokaryotes *in situ* (Brock, 1987). These methods commonly concentrate on the ability to extract total nucleic acids from cells occupying a sample and employing biomarker-based techniques, such as DNA analysis (Head et al., 1998). The use of ribosomal gene sequences as indicators of bacterial diversity is most common, although other genes (including protein-coding genes) have also been used (Wintzingerode et al., 1997). Although 16S rRNA gene research is more widely conducted, there are various culture-independent techniques that target other genes involved in metabolic actions.

Culture-independent approaches that use the 16S rRNA gene as a marker fail to take into account any differences in cellular metabolic activity, or to ensure that ecologically significant microbes are analysed, rather than microbial cells that exist in dormant forms that do not contribute to the function of the environmental system (Ellis et al., 2003). Hence, by incorporating both culture-dependent and -independent strategies, it is possible to target cells that are only active in specific metabolic functions coded for targeted genes.

Culture-independent methods may detect species that are missed by plating, provided that the amplification efficiency is high enough. However, they are typically dependent on PCR and other molecular biological techniques. Several potential biases have been shown or are conceivable for the required extraction of community DNA, the PCR step, and other enzymatic reactions. Furthermore, cloning into vectors or separation of 16S ribosomal DNA (rDNA) by denaturing or temperature gradient electrophoresis has its own potential shortcomings, regarding accurate taxonomic separation (Wikner, 2003). 16S rRNA gene sequences can be used for identification in studies of bacterial phylogeny and taxonomy. There are several reasons why 16S rRNA is used, including its presence in almost all bacteria, where it frequently exists as a multi-gene family or operons. Second, the function of the 16S rRNA gene has remained constant over time, implying that random sequence changes are a more accurate measure of time (evolution). The 16S rRNA gene is also large enough for informatics purposes (1500 bp) (Abbott, 2007).

16S Metagenomic Analysis

Metagenomics refers to genetic-based techniques used to extract the total genomic DNA obtained from micro-organisms without prior culturing in the laboratory, which can represent the vast majority of micro-organisms on earth (Hugenholtz et al., 1998). The new generation of sequencing technology, with its ability to sequence thousands of organisms in parallel, has proved to be uniquely suited to this application. Metagenomics is a genetics-based technique, which can be extremely valuable in providing an alternate approach for studying the proportion of the microbial community that is inaccessible by conventional microbial techniques. It attempts to capture the total genomes present in a community of interest. This method builds on recent advances in microbial genomics, as well as gene amplification and cloning through the use of polymerase chain reaction (PCR) for the creation of metagenomic libraries (Schloss and Handelsman, 2003). Direct genomic cloning offers the opportunity to capture operons or genes encoding pathways that may direct the synthesis of complex molecules, such as antibiotics (Schloss and Handelsman, 2003). In addition, culture-independent genomic analysis does not require prior sequence information for primer design, which is normally required for PCR-based methods. DNA is isolated from a microbial community, followed by the construction of the DNA libraries using suitable vectors, such as plasmids, cosmids, fosmids, and bacterial artificial chromosome (BAC) vectors (Kimura, 2006).

The new era of metagenomics was ushered in by studies utilizing 16S rRNA as a phylogenetic marker of microbial taxa (Pace, 1997). The 16S gene occurs in all living organisms (with the notable exception of viruses) and represents more than 80% of total bacterial RNA. The 16S rRNA gene structure is composed of interspersed conserved and variable regions, making it ideal for PCR amplification and sequencing. Probes can be designed to hybridize to the conserved regions, allowing for amplification and sequencing of the variable regions. Focusing on a small part of the microbial genome can dramatically reduce the sequencing cost. This approach has been particularly effective in monitoring microbial community fluctuations in populations (Caporaso et al., 2011).

Conclusion

Dinoflagellates are a very large and diverse group of eukaryotic organisms that play a major role in the aquatic food webs of both freshwater and marine habitats. Some dinoflagellate species can produce toxins, thereby causing damage to fishery industries and public health. Bacterial assemblages found in the phycosphere of dinoflagellates may play an important role in regulating dinoflagellate toxin production. The interactions between bacteria and dinoflagellates have been shown to be highly variable and are sometimes specific. While several studies have suggested that bacteria–phytoplankton interactions have the potential to dramatically influence harmful algal bloom dynamics, little is known about how bacteria and phytoplankton communities interact at the species composition level.

At present, the precise association between bacteria and cultured dinoflagellates is still not well understood. Moreover, current estimates indicate that more than 99% of the micro-organisms present in many natural environments are not readily culturable and, therefore, are not accessible for biotechnological or basic research (Amann et. al., 1995). Technology allowing for access to the genomic DNA or RNA of micro-organisms directly from environmental samples without prior cultivation has opened new pathways, serving as an important step towards better understanding bacteria-dinoflagellate interactions.

References

Amann, R. I., Ludwig, W. & Schleifer, K. H. (1995). Phylogenetic identification and in situ detection of individual microbial cells without cultivation. *Microbiological Reviews, 59*(1), 143-169.

Ashton, M. A. (2003). The effect of temperature on the toxicity of the laboratory cultured dinoflagellate *Ostreopsis lenticularis* and its associated bacterial flora. Ph.D. University of Puerto Rico, Mayagüez.

Barbier, J. L. (2008). Culture-Independent Methods for Identifying Microbial Communities in Cheese. *Food. Microbiol*. pp. 839-848.

Berland, B. R., Bonin, D. J. & Maestrini, S. Y. (1970). Study of bacteria associated with marine algae in culture. III. Organic substrates supporting growth. *Mar. Biol.* (Berlin) 5:68-76.

Brock, T. D. (1987). The study of microorganisms *in situ*: progress and problems. *Symp. Soc. Gen. Microbiol.* 41: 1-17.

Caporaso, J. G., Paszkiewicz, K., Field, D., Knight, R. & Gilbert, J. A. (2011). The Western English Channel contains a persistent microbial seed bank. *Int. Soc. Microbiol. Eco. J.* 6: 1089-1093.

Daranas, A. H., Norte, M. & Fernández, J. J. (2001). Toxic marine microalgae. *Toxicon.* 39:1101-32.

Doucette, G. J. & Tricks, C. G. (1995). Characterization of bacteria associated with different isolates of *Alexandrium tamarense*. *In Lassus, P., Arzul G., Erad E., Gentien, P. & Marcaillou, C.* [Eds.] *Harmful Marine Algal Blooms*. Lavoisier, Paris, pp. 33–38.

Ellis, R. J., Morgan, P., Weightman, A. J. & Fry, J. C. (2003). Cultivation-dependent and independent approaches for determining bacterial diversity in heavy-metal contaminated soil. *Appl. Environ. Microbiol.* 69: 3223-3230.

Ferrier, M., Martin, J. L. & Rooney-Varga, J. N. (2002). Stimulation of *Alexandrium fundyense* growth by bacterial assemblages from the Bay of Fundy. *J. Appl. Microbiol.* 92:706-716.

Gallacher, S. & E. A. Smith. (1999). Bacteria and paralytic shellfish toxins. *Protist.* 150: 245-255.

Gallacher, S., Flynn, K. J., Franco, J. M., Brueggemann, E. E. & Hines, H. B. (1997). Evidence for production of paralytic shellfish toxins by bacteria associated with *Alexandrium* spp. (*Dinophyta*) in culture. *Appl. Environ. Microbiol.* 63: 239-45.

González, I., Tosteson, C., Hensley, V. & Tosteson, T. R. (1995). Associated bacteria and toxicity development in cultured *Ostreopsis lenticularis*., p. 451-456. *In Lassus, P., Arzul, G., Erard, E., Gentien, P., and Marcaillou C.* (ed.), *Harmful marine algal blooms*. Lavoiser, Paris.

Hallegraeff, G. A. (1993). A review of harmful algal blooms and their apparent global increase. *Phycologia* 32(2): 79-99.

Hallegraeff, G. M. (1995). Harmful algal blooms: a global overview., p. 1-22. *In G. M. Hallegraeff, D. M. Anderson, and A. D. Cembella* (ed.), Manual on harmful marine microalgae, vol. 33. UNESCO.

Head, I. M., Saunders, J. R. & Pickup, R. W. (1998). Microbial evolution, diversity, and ecology: a decade of ribosomal RNA analysis of uncultivated microorganism. *Microbiol. Ecol.* 35: 1-21.

Horner-Devine, M. C, Carney, K. M. & Bohannan B. J. M. (2003). An ecological perspective on bacterial biodiversity. *The Royal Society* [online] https://www.tanfordd.edu/bohannan/Media/HornerDevineetal04%20copy.pdf [cited 2 December 2021].

Hugenholtz, P., Goebel, B. M. & Pace, N. R. (1998). Impact of culture-independent studies on the emerging phylogenetic view of bacterial diversity. *J. Bacteriol.* 180: 4765–4774.

Irigoien, X., Flynn, K. J. & Harris, R. P. (2005). Phytoplankton blooms: a"loophol"inmicrozooplankton grazing impact? *J. Plank. Res.* 27(4): 313-321.

Ivone Vaz-Moreira, C. E. (2011). Culture-dependent and culture-independent diversity surveys target different bacteria: a case study in a freshwater sample. *Antonie van Leeuwenhoek J. Microbiol.* p. 245–257.

Janda, J. M. & Abbott, S. L. (2007). 16S rRNA gene sequencing for bacterial identification in the diagnostic laboratory: pluses, perils, and pitfalls. *Journal of Clinical Microbiology*, *45*(9), 2761-2764.

Kimura, N. (2006). Metagenomics: Access to unculturable microbes in the environment. *Microbes. Environ.* 21(4):201-215.

Konopka, A. (2009). What is microbial community ecology? *Int. Soc. Microb. Ecol.* p. 1223–1230.

Landry, M. R. & Kirchman, D. L. (2002). Microbial community structure and variability in the tropical Pacific. *Deep Sea Research Part II: Topical Studies in Oceanography* 49(13-14): 2669-2693.

Leung, K. M. Y. & Dudgeon, D. (2008). Ecological risk assessment and management of exotic organisms associated with aquaculture activities. In *Understanding and applying risk analysis in aquaculture* (M. G. Bondad-Reantaso, J. R. Arthur, R. P. Subasinghe editors), FAO, Rome.

Lovejoy, C., Bowman, J. P. & Hallegraeff, G. M. (1998). Algicidal effects of a novel marine *Psedoalteromonas* isolate (class *Proteobacteria*, Gamma subdivision) on harmful algal bloom species of the genera *Chattonella*, *Gymnodinium*, and *Heterosigma*. *Appl. Envion. Microbiol.* 64: 2806-2813.

Pace, N. R. (1997). A molecular view of microbial diversity and the biosphere. *Science* 276: 734-740

Parab, S. G., Matondkar, S. G. P., Gomes, H. D. & Goes, J. I. (2006). Monsoon driven changes in phytoplankton populations in the eastern Arabian Sea as revealed by microscopy and HPLC pigment analysis. *Continental Shelf Research* 26(20): 2538-2558.

Pauly, D. (2003). Ecosystem impacts of the world's marine fisheries. *Global Change Newsletter*, 55:21.

Popovich, C. A., Spetter, C. V., Marcovecchio, J. E. & Freije, R. H. (2008). Dissolved nutrient availability during winter diatom bloom in a turbid and shallow estuary (Bahia Blanca, Argentina). *J. Coast. Res.* 24(1): 95-102.

Ryota Ishino, S. I. (2012). Bacterial Diversity Associated with the Rotifer *Brachionus plicatilis* sp. Complex Determined by Culture-Dependent and Culture-Independent Methods. *Biocontrol Science*. p. 51-56.

Schloss, P. D. & Handelsman J. (2003). Biotechnological prospects from metagenomics. *Curr. Opin. Biotech*. 14:303-310.

Smayda, T. J. (1997). What is a bloom? A commentary. *Limnol. Ocean.* 42(5): 1132- 1136.

Staley, J. T. & Konopka, A. (1985). Measurement of in situ activities of nonphotosynthetic microorganisms in aquatic and terrestrial habitats. *Annual. Rev. Microbiol.* 39: 321-346

Tang, Y. Z. & Gobler, C. J. (2009). Characterization of the toxicity of *Cochlodinium polykrikoides* isolates from Northeast US estuaries to finfish and shellfish. *Harmful Algae* 8(3): 454-462.

Thomson, P., Mcminn, A., Kiessling, I., Watson, M. & Goldsworthy, P. (2006). Composition and succession of dinoflagellates and chrysophytes in the upper fast ice of Davis Station, East Antarctica. *Polar Biology*. 29(4): 337.

Tosteson, T. R., Ballantine, D. L. & Winter, A. (1998). Sea surface temperature, benthic dinoflagellate toxicity, and toxin production transmission in the ciguatera food web, p. 48-49. In B. Reguera, J. Blanco, M. L. F. Pernández, and T. Wyatt (ed.), *Harmful algae*. Xunta de Galicia and intergovernmental oceanographic commission of UNESCO, Spain.

Van Dolah, F. M. (2000). Marine algal toxins: Origins, health effects, and their increased occurrence. *Environ. Health. Pers.* 108: 133-141.

Wear, R. G. & Gardner, J. P. A. (2001). Biological effects of the toxic algal bloom of February and March 1998 on the benthos of Wellington Harbour, New Zealand. *Mar. Ecol. Prog. Ser.* 218: 63-76.

Wikner, V. K. (2003). Combining Culture-Dependent and -Independent Methodologies for Estimation of Richness of Estuarine Bacterioplankton Consuming Riverine Dissolved Organic Matter. *Appl. Environ. Microbiol.*: 3607–3616.

Wintzingerode, F. V., Goebel, U. B. & Stackebrandt, E. (1997). Determination of microbial diversity in environmental samples: pitfalls of PCR-based rRNA analysis. *FEMS Microbiol. Rev.* 21: 213-229.

Yentsch, C. S., Lapointe, B. E., Poulton, N. & Phinney, D. A. (2008). Anatomy of a red tide bloom off the southwest coast of Florida. *Harmful Algae*. 7(6): 817-826.

Chapter 4

The Impact of Mercury Contamination in Seafood

Wen Jye Mok*
Siti Zuraida Zulkipli
and Wen Hui Zhang
Institute of Climate Adaptation and Marine Biotechnology, Universiti Malaysia Terengganu, Kuala Nerus, Terengganu, Malaysia

Abstract

Mercury naturally occurs on the earth's crust in a small quantity, exists in various forms; as inorganic mercury including metallic mercury and mercury vapor (Hg^0); and mercurous (Hg_2^{++}) or mercuric (Hg^{++}) salt; and organic mercury that is bonded with carbon atoms such as methyl, ethyl, phenyl, among others. Anthropogenic activities such as gold mining and smelting, urbanization, industrial and coal burning activities could lead to higher pollution levels. According to the UNEP report, East and Southeast Asia generated 39% of all atmospheric mercury emissions in 2015, compared to 0.4% from Australia, New Zealand, and Oceania. Bioaccumulation of mercury in fishes depend on the age, the amount of food eaten and the mercury concentration of their living environment throughout the life span with the concentration of mercury increases through biomagnification from the trophic level in the food web. Biomonitoring of mercury in seafood from wild or aquaculture organisms could be conducted with reasonable confidence about their validity through appropriate sample collection data; sampling date, location, species, total weight, total length and tissue analysed. The

* Corresponding Author's Email: mok.jye@umt.edu.my.

In: Climate Change and Environmental Perturbations
Editors: T. Min Pau, M. Abd. Ghaffar, M. Danish-Daniel Abdullah et al.
ISBN: 979-8-89113-271-9

thermal decomposition, amalgamation, and atomic absorption spectrophotometry of a Direct Mercury Analyzer, cold vapour atomic absorption spectrophotometer (CVAAS), electrothermal atomic absorption (ETAAS), and flame atomic absorption spectroscopy (FAAS) are the typical methods to analyse the total mercury in the specimen tissues. From 2007 to 2021, mercury contamination in Malaysian fish and seafood from aquaculture, capture fisheries, freshwater, marine, and mangrove environments was reported in the liver and muscle of the various species. According to standards established by the Joint FAO/WHO Expert Committee on Food Additives, total mercury concentrations in Malaysian seafood, specifically fish, were less than 0.5 g/g, which is within the permissible limit of mercury contamination in seafood.

Keywords: mercury analysis, mercury, seafood

Introduction

Characteristics of Mercury

Mercury (Hg) is an element with atomic number of 80 and atomic mass of 200.59. It is the only metal that is in liquid form and able to evaporate at room temperature. Its melting point is −38.87 °C (−37.967 °F), its boiling point is 356.58 °C (673.84 °F), and its density is 13.546 g/cm^3. It creates soft alloys (amalgams) when combined with other metals and has been widely used for plating. It is still used in thermometers and mercury vapor lamps. It should be highlighted that, while this element is easy to work with, it is highly toxic and harmful to humans if proper precautionary measures are not taken (Yorifuji, 2012).

Mercury naturally occurs in the Earth's crust in a small quantity (approximately 0.5 ppm) (Bernhoft, 2012). It may be exposed to the atmosphere due to outgassing from rocks or volcanic areas (Mok et al., 2011; Bernhoft, 2012). Mercury occurs in the earth's crust as native mercury and in mercury-bearing minerals, most commonly as toxic mercury sulfide (HgS), commonly known as cinnabar (Kidd & Batchelar, 2011). However, anthropogenic activities, such as gold mining and smelting, urbanization, and industrial and coal burning activities, could further increase mercury pollution levels (Kidd & Batchelar, 2011; Bernhoft, 2012). Mercury exists in various forms: as inorganic mercury, including metallic mercury, mercury vapor

(Hg^0), and mercurous (Hg_2^{++}) or mercuric (Hg^{++}) salts, and organic mercury bonded with carbon atoms, such as methyl, ethyl, or phenyl groups, among others (Bernhoft, 2012).

Mercury Pollution Cycle

The massive production of mercury from anthropogenic activities occurs in the form of elemental or inorganic metal, as wet or dry particles either evaporating into the atmosphere, carried by wind, or directly flowing into the aquatic environment (Strode et. al., 2007). The total emissions of mercury and releases from volcanic activity, biomass burning, soil, and vegetation, as well as anthropogenic activities, was estimated at 4400 tonnes in 2018, the highest record of mercury emission into the atmosphere to date (Figure 1) (UNEP, 2018a). Relevant anthropogenic activities include fuel combustion, industrial activity, and intentional uses (e.g., product wastes and artisanal and small-scale gold mining). East and Southeast Asia contributed the highest atmospheric mercury emissions at 39% in 2015, while Australia, New Zealand, and Oceania only contributed 0.4% (UNEP, 2018a). According to Habuer et al., (2016), the major sources of mercury emissions in Malaysia come from the solid waste incineration facilities (7.8–35.1 $\mu g/Nm^3$) and coal-fired power stations (8.5 $\mu g/Nm^3$). The gaseous mercury in the atmosphere turns into rain and is permanently deposited into aquatic systems. Others may be directly deposited into the sea from wastewater or soil due to anthropogenic activities. The inorganic mercury will then settle at the seabed, while bacteria methylate the mercury and change it from an inorganic to organic form—namely, methylmercury (MeHg)—which easily binds to the tissues of living organisms. The major bacterial group that has been identified as mercury-methylating bacteria is sulfate-reducing bacteria (Parks et al., 2013). MeHg bioaccumulates and biomagnifies in the environment and food web.

The bioaccumulation of mercury in fishes depends on their age, the amount of food eaten, and the concentration of mercury in the living environment throughout their life span. The concentration of mercury increases through biomagnification from the trophic level in the food web, starting from primary producers to predatory fish. Thus, the top predators in a food web may have concentrations of MeHg in their tissues higher than the concentrations in the living environment (UNEP, 2018). However, areas with high mercury deposition rates do not necessarily have high MeHg concentrations in the organisms of the food web, or vice versa. For example, atmospheric mercury wet deposition levels measured in the United States and southern Canada were the lowest, but high MeHg concentrations have been reported in fish and birds in North America (UNEP, 2018). The dietary consumption of fish, shellfish, marine mammals, and other foods is considered the most important source of MeHg exposure in human. Thus, biomonitoring of mercury in seafood should be undertaken in the region for assessment and monitoring of the risks posed by mercury to ecology and human health.

History of Minamata Disease

Mercury pollution has been of concern since the 1950s due to the outbreak of Minamata disease in Japan, with symptoms and signs of a disorder affecting the central nervous system (Saito, 2020). This poisoning caused serious health problems among the residents of Minamata City, Kumamoto Prefecture. The second poisoning reported was in Niigata, at a factory located in Kanose Town about 65 km from the mouth of the Agona River, which produced acetaldehyde and released MeHg as a by-product into the environment (Saito et al., 2020). Both cases were due to the discharge of mercury into the environment from the factories which polluted the sea and river nearby, causing poisoning from consumption of fish contaminated with MeHg levels as high as 40 mg/kg (Saito et al., 2020). High concentrations of mercury—above 50 mg/kg with the highest being 705 mg/kg—were detected in the hair of patients (Saito, et al., 2020; Fujiki & Tajima, 1992). The characteristic symptoms of Minamata disease include intellectual impairment, persistence of primitive reflexes (e.g., sucking, rooting, grasping, and crossed extension), cerebellar signs, delayed physical development in children, reproduction, behaviour, and survival (UNEP, 2018; Saito, 2020).

Method for Mercury Analysis in Seafood

Sample Collection and Preparation

The biomonitoring of mercury in seafood from wild or aquaculture organisms can be conducted with reasonable confidence regarding its validity through appropriate sample collection data; for example, considering the sampling date, location, species, total weight, total length, and tissue analysed.

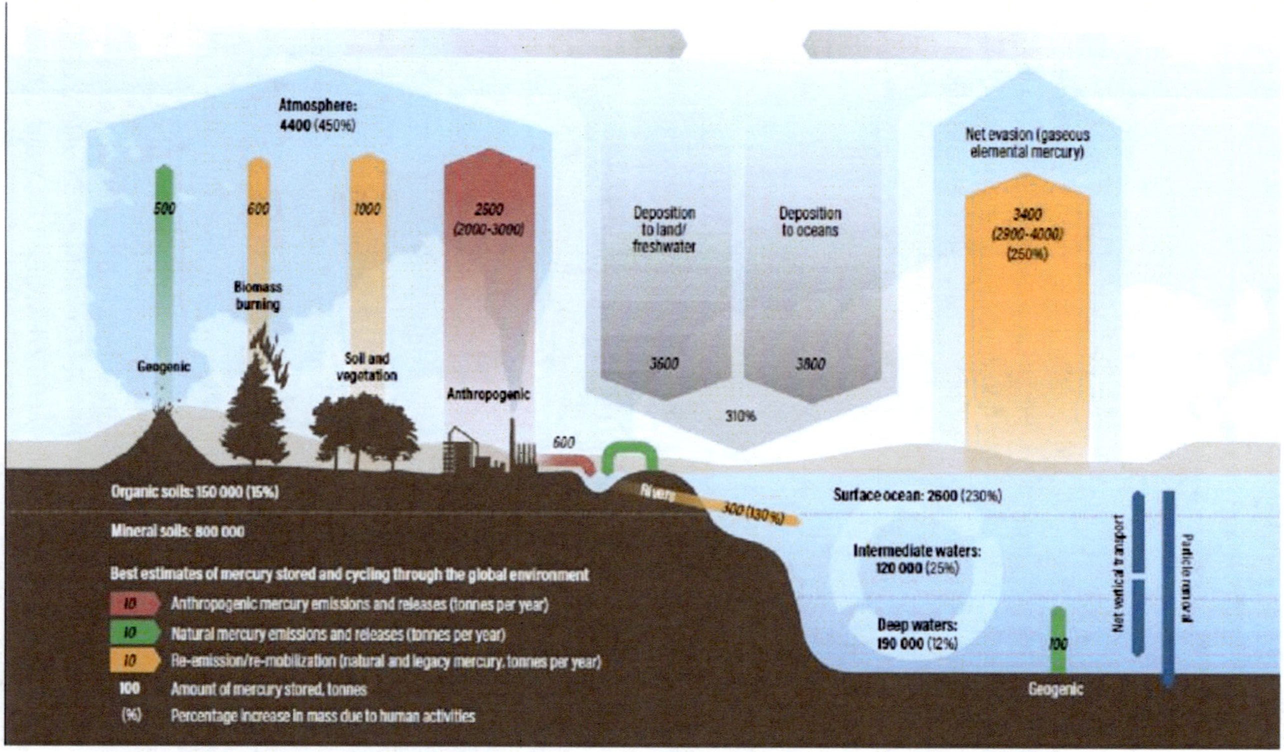

Figure 1. Global Hg emissions from natural and anthropogenic activities in the Hg cycle and the Hg accumulated in the atmosphere, soils, and oceans (UNEP, 2018).

The Ministry of the Environment of Japan has suggested that, for fish, about 10–20 grams of the edible parts should be collected and placed in polyethylene bags or containers, then stored in a freezer in the laboratory. For shellfish samples, the muscle, digestive tract contents, and adductor muscle (snails without an adductor muscle should only be divided into the edible portion) are divided and the portions placed in separate polyethylene bags or containers, then stored in a freezer. All sediment particles in the digestive tracts of shellfish need to be removed before storage. This is followed by sample preparation before analysis using the appropriate instruments. There are several ways to carry out sample preparation, such as conventional acid digestion or wet digestion. At present, these digestion procedures can be performed using microwave digestion. Other methods of sample preparation are referred to in Table 1.

Direct Mercury Analyser

There are several ways to measure the total mercury concentration in solid samples. One of the most common methods is to analyse the total mercury in the specimen tissues based on the US EPA method 7473 (USEPA, 2007), through thermal decomposition, amalgamation, and atomic absorption spectrophotometry using a Direct Mercury Analyser (DMA-80 evo; Milestone, Sorisole, Italy). This instrument includes a heating-controlled oxygenated decomposition furnace to liberate mercury from the samples. The decomposition products are then carried to an amalgamator to trap the mercury while the system is flushed with oxygen to remove all remaining gases or decomposition products. The amalgamator is rapidly heated to release mercury vapor, where the flowing oxygen will carry the mercury vapor through absorbance cells into the light path of a single-wavelength atomic absorption spectrophotometer, allowing it to be measured at 253.7 nm.

For quality control, a calibration check with at least triplicates of the blank (run without nickel boats) and triplicates of certified reference materials (CRM)—namely, SRM2976 (mussel; National Institute of Standard & Technology, USA) with mercury concentration 61.0 (± 3.6) μg/kg of biota sample—is the standard protocol. Recovery of SRM2976 used for calibration should be within 95–99% of the certified value. Once the calibration is completed, about 10–20 mg of dried (dried using freeze dryer) or wet (untreated) tissue samples is weighed in a nickel boat and placed in an autosampler for the total mercury analysis. There are several other CRMs that could be used for total mercury or speciation of mercury in marine animal tissues other than SRM2976, such as ERM-CE464 (tuna fish muscle), SRM 1566 b (oyster tissue), SRM 1974 B or SRM 2977 (mussel tissue), DORM-2 (dogfish muscle), DORM-3 (fish protein), DOLT-3 or DOLT-4 (dogfish liver), LUTS-1 (non-defatted lobster tissue), and TORT-2 (lobster tissue). The selection of CRM depends on the user's preferences regarding the instrument, limit of detection, and availability.

Other Methods and Instruments for Mercury Analysis in Seafood

In addition to the method of thermal oxygenated decomposition, amalgamation, and atomic absorption spectrophotometry (using a direct mercury analyser), several approaches can be used to analyse mercury or mercury speciation concentration, such as cold vapor atomic absorption spectrophotometry (CVAAS), electrothermal atomic absorption (ETAAS), flame atomic absorption spectroscopy (FAAS), and so on, as detailed in Table 1. There are also several standard operation protocols for the preparation of samples and analyses using different instruments. For example, EPA methods 245.1, 245.2, 7470A, and 7471A are suitable for CVAAS; EPA methods 200.8 for ICP-MS; EPA method 1631 E and 245.7 for CV-AFS; EPA methods 6800 for SIDMS; and so on. Each instrument has certain limitations for mercury analysis, depending on the speciation of mercury, sensitivity of detection, limit of detection, recovery, sample preparation, and cost of operation. Almost all instruments require sample preparation before analysis, except when using a Direct Mercury Analyser, which makes analysis direct and easy while maintaining a low limit of detection and high sensitivity and recovery. Thus, for the testing of total mercury, it is suggested to use a Direct Mercury Analyser for ease and simplicity of operation.

Seafood In Malaysia

Seafood Production Statistics in Malaysia

Seafood is considered one of the most important protein sources in Malaysia. It is rich in unsaturated essential fatty acids—especially eicosapentaenoic acid

(EPA) and docosahexaenoic acid (DHA)—as well as minerals and vitamins (Burger & Gochfeld, 2009). As such, it has beneficial effects on human health, such as reducing cholesterol levels, heart disease, stroke, and pre-term delivery (Ahmad, 2016). According to the Department of Fisheries (DOF) (2021), the production of captured fisheries and aquaculture from 2015 to 2020 averaged 1.89 million metric tonnes, with capture fisheries amounting to 1.47 million metric tonnes and aquaculture production yielding 0.42 million metric tonnes (Figure 2). Ahmad et. al. (2016) has stated that the average fish consumption for a Malaysian is currently about 168 g/day or 54.77 kg/person/year, leading Malaysia to rank third in fish consumption worldwide, after only Japan (64.7 kg/person/year) and Korea (51 kg/person/year). The most-consumed fish in Malaysia is the Indian mackerel, followed by anchovy, yellowtail, yellow-stripe scads, tuna, sardines, torpedo scads, Indian and short-fin scads, pomfret, red snapper, king mackerel, threadfin bream, croaker, marine catfish, stingray, barramundi, and so on (Ahmad et al., 2016).

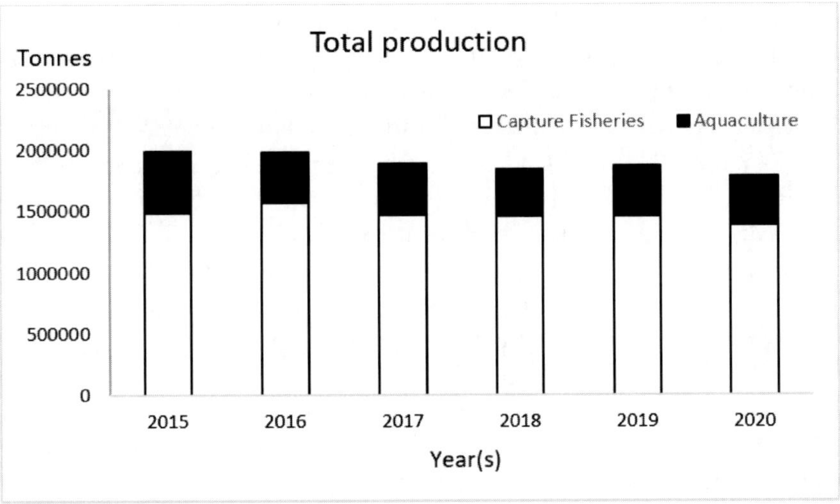

Figure 2. The total capture from fisheries and aquaculture production from 2015 to 2020 in Malaysia (DOF).

Table 1. Methods and instruments used for analysis of mercury in aquatic organisms/seafood

No.	Analysed samples	Analyte	Sample preparation	Analytical instrument	Reference
1	Crab, prawn, abalone, whitebait, snapper, bass, oyster, squid, dace	Total Hg	None	Thermal oxygenated decomposition, amalgamation, atomic absorption spectrophotometer (Direct Mercury Analyser)	Chen et al., 2005
2	Fish tissue; Canned tuna fish; Fish samples; Fish muscles; Mussel Tissue and Lobster Hepatopancreas	Total Hg	Wet digestion or acid digestion	Cold vapor atomic absorption spectrophotometer (CVAAS)	Chang et al., 2016; Elisabeth et al., 2018; Rahimi et al., 2010; Ramezani et al., 2010; Miklavi et al., 2011; Molaei et al., 2011; Romera et al., 2011; Voegborlo & Adimado; 2010
3	Fish tissue	Total Hg	Single-drop microextraction	Electrothermal atomic absorption (ETAAS)	Martinis & Wuilloud, 2010
4	Fish samples	Total Hg	Acid digestion	Flame atomic absorption spectroscopy (FAAS)	Xiao, 2004; Yamashita et al., 2011
5	Dogfish muscle	Hg speciation	Photo-induced chemical vapor generation	Inductively coupled cold vapor atomic fluorescence spectrometry (IC-CV AFS)	Liu, 2010
6	Fish	Hg and MeHg	Amalgamation	Cold vapor atomic fluorescence spectrometry (CV AFS)	Williams et al., 2011
7	Shrimp and fish	Hg	Microwave digestion	Hydride generation atomic absorption spectrometry (AAS-FIAS)	Gomez & Cartagena, 2014
8	Canned crab meat, large yellow croaker, mud clam, crucian carp, shrimp, river crab	Hg speciation	Ultrasound-assisted extraction	High-performance liquid chromatography atomic fluorescence spectrometry (HPLC-AFS)	Fan et al., 2017
9	Tuna fish	MeHg	Microwave digestion	Inductively coupled plasma mass spectrometry (ICP-MS)	Reyes et al., 2008
10	Fish samples	Hg	Microwave digestion	Inductively coupled plasma optical emission spectrometry (ICP-OES)	Ural et al., 2010

Mercury Concentration in Seafood

Mercury is ranked, by the U.S. Government Agency for Toxic Substances and Disease Registry, as the third-most toxic element on the planet, after arsenic (As) and lead (Pb), which are continuously being dumped into waterways and soil, spilled into the atmosphere, and polluting food and water sources (Rice et al., 2014), as discussed earlier. Consumption of fish or seafood is the major route of exposure to mercury among consumers. Most consumers know that eating fish is beneficial to their health, but less may be aware of the mercury contamination in fish that could post risks to human health, as it can cause neuron dysfunction. Burger & Gochfeld (2009) have stated that less than 25% of the 329 respondents in their research mentioned the risk of mercury in fish, revealing that the issue of potential risk of fish consumption may not be widely known to the public. The World Health Organization (WHO) recommends a safe intake of MeHg of no more than 1.6 μg kg-1 per week (FAO, 2003). The United States Environmental Protection Agency (EPA) and National Research Council (NRC) developed a safe reference dose of 0.1 μg kg-1 body weight per day for adults (NRC, 2000).

Table 2 represents the mercury contamination detected in the liver and muscle in dry and wet weights of different species of fish or seafood from captured fisheries and aquaculture, as well as freshwater, seawater, and mangrove areas in Malaysia from 2007 until 2021 (Agusa et al., 2007; Hajeb et al., 2009; Mok et al., 2012; Jeevanaraj et al., 2016; Il et al., 2018, Siau et al., 2021). The total mercury concentrations found in these seafood—specifically, in fish—were less than 0.5 μg/g, which is within the safe limit of mercury contamination in seafood based on the guidelines established by the Joint FAO/WHO Expert Committee on Food Additives (FAO, 2016). The amount of permitted mercury in predatory fish is no more than 1 μg/g (FAO, 2016). Agusa et al., (2007) have shown that the mercury concentrations in the liver and muscle of all fish species investigated in four states comprising of Terengganu, Johor, Negeri Sembilan, and Kedah were lower than 0.5 μg/g, except for the fork-tailed threadfin bream (Terengganu) with mercury contents of 2.40 μg/g dry weight in liver and 0.67 μg/g dry weight in muscle, exceeding the safe 0.5 μg/g limit (Table 2). Basically, the mercury concentrations in dry weight conditions are expected to be significantly higher than those in wet weight conditions (Hajeb et al., 2009). If there is a considerable 80% loss of moisture in the dried samples (Alonso-Fernández & Saborido-Rey, 2012), the concentration in the muscle for fork-tailed threadfin bream might be still under 0.5 μg/g. Nevertheless, depending on how the fish is served to consumers (i.e., whether it is cooked with wet weight or served as dried fish), its safety is uncertain. Other studies have shown that the total mercury concentrations in seafood in Malaysia were still within the tolerable limit of 0.5 μg/g (Hajeb et al., 2009; Mok et al., 2012; Jeevanaraj et al., 2016; Anual et al., 2018; Siau et al., 2021).

The different aquatic biomes of biological resources also influence the level of mercury content. Freshwater fish (Mok et al., 2012) had significantly lower mercury concentrations in their muscle, compared to marine (Jeevanaraj et al., 2016) and mangrove (Siau et al., 2021) fishes, which could be attributed to the mercury concentration levels in their respective living environments. Praveena et al., (2013) reported that high concentrations of Hg in the areas of West Port, Malacca Straits, Prai, and Johor were due to industrial activities, where the total mercury concentration in sediments from West Port and Sungai Pulau, Johor ranged from 0.11 to 0.41 mg/kg, while levels of 0.03 to 0.08 mg/kg were detected in Sungai Pok Kecil, Sungai Pok Besar, and Teluk Buih in Johor, as well as around Merambong Island. However, the mercury concentration in the sediment around the mangrove areas in Peninsular Malaysia ranged from 2.66–311.63 μg/kg (Cheng & Yap, 2015). The mercury concentration found in crustaceans—specifically, marine shrimp—also showed significantly lower mercury concentration in muscle than that in marine fish species (Mok et al., 2012). This is closely related to the bioaccumulation and biomagnification in the water column and food chain, as discussed previously.

There are still many fish species that have yet to be tested for their mercury concentration, whether total mercury, inorganic mercury, or MeHg, especially predatory species. Comprehensive studies on environmental mercury contamination should be carried out, as the UNEP has marked Southeast Asia as a major regional source of mercury emission. Periodical analysis should be conducted for at least total mercury concentration on those widely consumed species, in order to ensure that these species are not becoming highly contaminated with mercury.

Table 2. Studies on seafood mercury contamination in Malaysia

Local name	Common name	Species	Location	Aquatic Biome	Mercury concentration in liver (µg/g)		Mercury concentration in muscle (µg/g)		Reference
					Dry weight	Wet weight	Dry weight	Wet weight	
Kerisi Ros Merah	Fork-Tailed Threadfin Bream	*Nemipterus furcosus*	Cabang Tiga, Terengganu	Marine	2.40		0.67		Agusa et al., 2007
Aya Kurik	Kawakawa	*Euthynnus affinis*		Marine	0.10		0.09		
Cencaru	Torpedo Scad	*Megalaspis cordyla*		Marine	0.15		0.21		
Lolong Mata Besar	Bigeye Scad	*Selar crumenophthalmus*	Kuala Terengganu, Terengganu	Marine	0.12		0.10		
Selayang Ekor Merah	Redtail Scad	*Decapterus kurroides*		Marine	0.15		0.16		
Lolong Mata Besar	Bigeye Scad	*Selar crumenophthalmus*	Mersing, Johor	Marine	0.26		0.36		
Kembung Borek	Indian Mackerel	*Rastrelliger kanagurta*		Marine	0.12		<0.05		
Ambah	Javelin Grunter	*Pomadasys kaakan*		Marine	0.06		0.09		
Lolong Mata Besar	Bigeye Scad	*Selar crumenophthalmus*	Parit Jawa, Johor	Marine	0.12		0.11		

Local name	Common name	Species	Location	Aquatic Biome	Mercury concentration in liver (µg/g)		Mercury concentration in muscle (µg/g)		Reference
					Dry weight	Wet weight	Dry weight	Wet weight	
Talang Bintik Lapis	Doublespotted Queenfish	*Scomberoides lysan*		Marine	0.36		0.09		
Lolong Mata Besar	Bigeye Scad	*Selar crumenophthalmu s*	Port Dickson, Negeri Sembilan	Marine	<0.05		<0.05		
Bawal Hitam	Black Pomfret	*Parastromateus niger*		Marine	0.21		<0.05		
Cencaru	Torpedo Scad	*Megalaspis cordyla*		Marine	0.18		0.27		
Lolong Mata Besar	Bigeye Scad	*Selar crumenophthalmu s*	Langkawi, Kedah	Marine	<0.05		<0.05		
Lidah Lumpur	Doublelined Tonguesole	*Paraplagusia bilineata*		Marine	0.24		0.09		
Selayang Ekor Merah	Redtail Scad	*Decapterus kurroides*		Marine	<0.05		0.05		
Basung	Shortfin Scad	*Decapterus macrosoma*		Marine	<0.05		0.06		
Cencaru	Torpedo Scad	*Megalaspis cordyla*		Marine	0.09		0.10		
Bandan	Yellowfin Seabream	*Acanthopagrus latus*		Marine	0.35		0.31		
Kembong	Short-Bodied Mackerel	*Rastrelliger brachsoma*	Selangor	Marine	0.03	0.02	0.45	0.15	Hajeb et al., 2009
Belata	Blackfin Scad	*Alepes vari*		Marine	0.02	0.01	0.04	0.00	

Table 2. (Continued)

Local name	Common name	Species	Location	Aquatic Biome	Mercury concentration in liver (μg/g)		Mercury concentration in muscle (μg/g)		Reference
					Dry weight	Wet weight	Dry weight	Wet weight	
Tenggiri	Narrow-Bodied Spanish Mackerel	*Scomberomorous commerson*		Marine	0.05	0.03	0.04	0.01	
Bawal Hitam	Black Pomfret	*Parastromateus niger*		Marine	0.10	0.06	0.13	0.04	
Aya Hitam	Long Tail Tuna	*Thunus Tonggol*		Marine	0.49	0.12	0.50	0.12	
Kerapu Bunga	Greasy Grouper	*Epinephelus tauvina*		Marine	0.03	0.01	0.03	0.01	
Kebasi Kuasi	Chacunda Gizzard Shad	*Anodontostoma chacunda*		Marine	0.03	0.01	0.09	0.04	
Selar Kuning	Yellow-Banded Scad	*Selaroides leptolepis*		Marine	0.04	0.03	0.06	0.03	
Aya Kurik	Kawakawa	*Euthynnus afinis*		Marine	0.02	0.01	0.05	0.01	
Gerisi Merah	Delagoa Threadfin Bream	*Nemipterus bipunctatus*		Marine	0.09	0.03	0.09	0.03	
Bukai-Bukai	Asian Seabass	*Lates calcarifer*		Marine	0.01	0.01	0.10	0.02	
Basung	Blue Mackerel Scad	*Decapterus macrosoma*		Marine	0.01	0.01	0.00	0.00	
Kerapu Bintang	Coral Trout	*Plectropomus leopardus*	Sabah	Marine		0.22		0.09	Mok et al., 2012

Local name	Common name	Species	Location	Aquatic Biome	Mercury concentration in liver (μg/g)		Mercury concentration in muscle (μg/g)		Reference
					Dry weight	Wet weight	Dry weight	Wet weight	
Benung	Fourfinger Threadfin	*Eleutheronema tetradactylum*	Johor	Marine		0.04		0.09	
Bawal	Pomfret	*Brama brama*	Sabah	Marine		0.15		0.16	
Bukai-Bukai	Asian Seabass	*Lates calcarifer*	Selangor; Negeri Sembilan; Pahang; Johor; Sabah	Marine		0.06		0.07	
Kerapu Harimau	Tiger Grouper	*Epinephelus fuscoguttatus*	Sabah	Marine		0.23		0.13	
Keli	African Catfish	*Clarias gariepinus*	Selangor; Negeri Sembilan; Pahang; Sarawak	Freshwater		0.10		0.06	
Puyu Kukum	Jade Perch	*Scortum Barcoo*	Selangor	Freshwater		0.14		0.08	
Jelawat	Mad Barb	*Leptobarbus hoevenii*	Sarawak	Freshwater		0.10		0.06	
Patin	Striped Catfish	*Pangasius hypophthalmus*		Freshwater		0.22		0.10	
Talapia	Tilapia	*Niloticus niloticus*	Selangor; Negeri Sembilan; Perak; Pahang; Sarawak	Freshwater		0.09		0.04	

Table 2. (Continued)

Local name	Common name	Species	Location	Aquatic Biome	Mercury concentration in liver (µg/g)		Mercury concentration in muscle (µg/g)		Reference
					Dry weight	Wet weight	Dry weight	Wet weight	
Udang Harimau	Tiger Prawn	*Penaeus monodon*	Selangor; Negeri Sembilan; Sabah;	Marine crustacean	-			0.06	
Udang Putih	Whiteleg Prawn	*Penaeus vannamei*	Selangor; Negeri Sembilan; Perak	Marine crustacean	-			0.04	
Basung	Blue Mackerel Scad	*Decapterus macrosoma*	Selangor	Marine			0.27		Jeevanaraj et al., 2016
Andapong	Flathead Grey Mullet	*Mugil cephalus*		Marine			0.28		
Bawal Hitam	Black Pomfret	*Parastromateus niger*		Marine			0.26		
Bawal Putih	Silver Promfret	*Pampus argentus*		Marine			0.24		
Bebulus	Silver Whiting	*Sillago sihama*		Marine			0.32		
Kembung Borek	Indian Mackerel	*Rastrelliger kanagurta*		Marine			0.27		
Selar Kuning	Yellow-Banded Scad	*Selaroides leptolepis*		Marine			0.27		
Cencaru	Torpedo Scad	*Megalaspis cordyla*		Marine			0.29		

Local name	Common name	Species	Location	Aquatic Biome	Mercury concentration in liver (µg/g)		Mercury concentration in muscle (µg/g)		Reference
					Dry weight	Wet weight	Dry weight	Wet weight	
Cabuk	Dorab Wolf Herring	*Chirocentrus dorab*		Marine			0.38		
Gubal	Bamin	*Eleutheronema tetradactylum*		Marine			0.34		
Tenggiri Papan	Indo-Pacific King Mackerel	*Scomberomorus guttatus*		Marine			0.37		
Aya Kurik	Eastern Little Tuna	*Euthynus affinis*		Marine			0.36		
Pari	Pale-Edged Stingray	*Dasyatis zugei*		Marine			0.42		
Gelama Janggut	Bearded Jewfish	*Johnius dussumieri*		Marine			0.31		
Barah-Barahan	Golden Snapper	*Lutjanus johnii*		Marine			0.48		
Kerapu Bintik Bulat	Areolate Grouper	*Epinephelus areolatus*		Marine			0.31		
Kerisi	Japenese Threadfin Bream	*Nemipterus japonicus*		Marine			0.36		
Ikan Merah	Red Snapper	*Lutjanus argentimaculatus*		Marine			0.37		
Bukai-Bukai	Giant Perch	*Lates calcarifer*		Marine			0.25		
Lolong Kuning	Oxeye Scad	*Selar boops*	West Coast of Peninsular	Marine				0.20	Anual et al., 2018
Cencaru	Torpedo Scad	*Megalaspis cordyla*		Marine				0.21	

Table 2. (Continued)

Local name	Common name	Species	Location	Aquatic Biome	Mercury concentration in liver (μg/g)		Mercury concentration in muscle (μg/g)		Reference
					Dry weight	Wet weight	Dry weight	Wet weight	
Tenggiri Papan	Indo-Pacific King Mackerel	*Scomberomorus guttatus*		Marine				0.18	
Tuna Tenggiri	Dogtooth Tuna	*Gymnosarda unicolor*		Marine				0.61	
Ikan Merah	Mangrove Red Snapper	*Lutjanus argentimaculatus*		Marine				0.28	
Merah	Malabar Blood Snapper	*Lutjanus malabaricus*		Marine				0.13	
Bahab'	John's Snapper	*Lutjanus russellii*		Marine				0.52	
Coreng	Emperor Red Snapper	*Lutjanus sebae*		Marine				0.14	
Bukai-Bukai	Asian Seabass	*Lates calcarifer*		Marine				0.25	
Pari Lalat	Bluespotted Stingray	*Dasyatis kuhlii*		Marine				0.27	
Gelama	Soldier Croaker	*Nibea soldado*		Marine				0.16	
Gelama Selampai	Bronze Croaker	*Otolithoides biauritus*		Marine				0.19	
Kerisi	Yellowbelly Threadfin Bream	*Nemipterus bathybius*		Marine				0.32	
Kerisi	Japanese Threadfin Bream	*Nemipterus japonicus*		Marine				0.24	

Local name	Common name	Species	Location	Aquatic Biome	Mercury concentration in liver (µg/g)		Mercury concentration in muscle (µg/g)		Reference
					Dry weight	Wet weight	Dry weight	Wet weight	
Kerisi	Doublewhip Threadfin Bream	*Nemipterus nematophorus*		Marine				0.65	
Udang Merah Ros	Pink Shrimp	*Metapanaeus affinis*		Marine				0.12	
Sotong	Indian Squid	*Loligo duvaucelli*		Marine				0.22	
Seriding	Scalloped Perchlet	*Ambassis nalua*	Setiu Wetlands, Terengganu	Mangrove			0.41		Siau et al., 2021
Bagat	Bigeye Trevally	*Caranx sexfaciatus*		Mangrove			0.41		
Kerapu Bintik	Brown-Spotted Grouper	*Epinephelus coioides*		Mangrove			0.16		
Bukai-Bukai	Asian Seabass	*Lates calcarifer*		Mangrove			0.13		
Gedabang	Common Ponyfish	*Leiognathus equulus*		Mangrove			0.25		
Asuh-Asuh	Bridled Pigface-Bream	*Lethrinus lentjan*		Mangrove			0.21		
Janahak/ Bahaba	Fingermark Bream	*Lutjanus russellii*		Mangrove			0.18		
Gelama	Four Lined Therapon	*Pelates qudrilineatus*		Mangrove			0.12		

Conclusion

As mercury is considered the third-most toxic element and can be a long-term accumulated poison, periodical monitoring of environmental mercury exposure should be conducted and a better understanding of its bioaccumulation in organisms and food chains should be obtained. Regulations for the disposal of mercury into the environment, whether in laboratories, factories, and mining activities, should be strictly adhered to. The public are advised to consume more of the smaller-sized fish with shorter life spans than those of large size and with long life span, due to the bioaccumulation rate in fishes.

References

Ahmad, N. I., Mahiyuddin, W. R. W, Tengku Mohamad, T. R., Ling, C. Y., Daud, S. F., Hussein, N. C., Abdullah, N. A., Shaharudin, R. & Sulaiman, L. H. (2016). Fish consumption pattern among adults of different ethnics in Peninsular Malaysia. *Food & Nutrition Research*, 60, 32697. https://doi.org/10.3402/fnr.v60.32697.

Alonso-Fernández & Saborido-Rey. (2012). Relationship between energy allocation and reproductive strategy in *Trisopterus luscus*. *Journal of Experimental Marine Biology and Ecology*, 416-417, 8-16. https://doi.org/10.1016/j.jembe.2012.02.001.

Bernhoft, R. A. (2012). Mercury toxicity and treatment: A review of the literature. *Journal of Environmental and Public Health*, 2012, 460508. https://doi.org/10.1155/2012/460508.

Burger, J. & Gochfeld, M. (2009). Perceptions of the risks and benefits of fish consumption: Individual choices to reduce risk and increase health benefits. *Environmental Research*, 109, 343-349. https://doi.org/10.1016/j.envres.2008.12.002.

Chang, J., Zhang, X., Mei, G., Gao, X., Liu, W. & Han, C. (2016). Determination of Total Mercury in Aquatic Products by Cold Atomic Absorption Spectrometry. *Journal of Zhejiang Ocean University (Natural Science)*, 4, 310-314.

Chen, C., Luo, Y. & Fan, Z. (2005). Fast determination of total mercury in aquatic products by thermal oxygenated decomposition-amalgamation-atomic absorption spectrophotometer. *Chinese Journal of Health Laboratory Technology*, 2005, 53-54.

Cheng, W. H. & Yap, C. K. (2015). Potential human health risks from toxic metals via mangrove snail consumption and their ecological risk assessments in the habitat sediment from Peninsular Malaysia. *Chemosphere*, 135, 156-165. https://doi.org/10.1016/j.chemosphere.2015.04.013.

DOF (Department of Fisheries Malaysia) (2021). *Annual Fisheries Statitistics* – Volume 1. https://www.dof.gov.my/sumber/perangkaan-perikanan-i/.

Elisabeth, M., Terry, M. & Azad, M. (2018). Optimization of instrument conditions for the analysis for mercury, arsenic, antimony and selenium by atomic absorption spectroscopy. *MethodsX,* 5, 824-833. https://doi.org/10.1016/j.mex.2018.07.016.

Fan, X., Zhang, R., Liu, B., Wang, M., Deng, X. & He, Y. (2017). Determination the content of mercury in different speciations in aquatic products by high performance liquid chromatography-atomic fluorescence spectrometry. *Journal of Food Safety and Quality*, 8, 76-81.

Fujiki, M. & Tajima, S. (1992). The pollution of Minamata Bay by mercury. *Water Science & Technology*, 25, 133-140. https://doi.org/10.2166/wst.1992.0284.

Habuer, Yoshimoto, N., Takaoka, M., Fujimori, T., Oshita, K. Sakai, N. & Abd Kdir, S. A. A. (2016). Substance flow analysis of mercury in Malaysia. *Atmospheric Pollution Research*, 7, 799-807. https://doi.org/10.1016/j.apr.2016.04.005.

Joint FAO/WHO Expert Committee on Food Additives (2003: Rome, Italy) Evaluation of certain food additives and contaminants: sixty-first report of the Joint FAO/WHO Expert Committee on Food Additives. https://apps.who.int/iris/bitstream/handle/10665/42849/WHO_TRS_922.pdf?sequence=1&isAllowed=y.

Kidd, K. & Batchelar, K. (2011). Mercury. *Fish Physiology*, 31, 237-295. https://doi.org/10.1016/S1546-5098 (11)31027-8.

Konieczka, P., Misztal-Szkudlińska, M., Namieśnik, J. & Szefer, P. (2010). Determination of total mercury in fish and cormorant using cold vapour atomic absorption spectrometry. *Polish Journal of Environmental Studies*, 19, 931-936.

Liu, Q. (2010). Determination of mercury and methylmercury in seafood by ion chromatography using photo-induced chemical vapor generation atomic fluorescence spectrometric detection. *Microchemical Journal*, 95(2), 255-258. https://doi.org/10.1016/j.microc.2009.12.010.

Martinis, E. M. & Wuilloud, R. G. (2010). Cold vapor ionic liquid-assisted headspace single-drop microextraction: a novel preconcentration technique for mercury species determination in complex matrix samples. *Journal of Analytical Atomic Spectrometry*, 25(9), 1432-1439.

Miklavi, A. Stibilj, V., Heath, E., Polak, T., Tratnik, J. S., Klav, J., Mazej, D. & Horvat, M. (2011). Mercury, selenium, pcbs and fatty acids in fresh and canned fish available on the Slovenian market. *Food Chemistry*, 124(3), 711-720. https://doi.org/10.1016/j.foodchem.2010.06.040.

Ministry of the Environment, Japan. 2004. Mercury analysis manual. http://nimd.env.go.jp/kenkyu/docs/march_mercury_analysis_manual(e).pdf.

Mok, W.J., Seoka, M., Tsukamasa, Y., Kawasaki, K. & Ando, M. (2011). Mercury levels of small fishes: influence of size and catch area. *Fisheries Science*, 77, 823-828. https://doi.org/10.1007/s12562-011-0392-1

Molaei, F., Nikpour, Y., Kamrani, E. & Carvalho, C. (2011). Determining mercury contamination in Indian spiny turbot, *Psettodes erumei*, from Jask Port coastal waters, Iran. *Journal of the Persian Gulf (Marine Science)*, 2, 43-48.

National Research Council (NRC), Toxicological Effects of Methylmercury, National Research Council, National Academy Press, Washington, DC, 2000.

Parks, J. M., Johs, A., Podar, M., Bridou, R., Hurt, R. A., Smith, S. D., Tomanicek, S. J., Qian, Y., Brown, S. D., Brandt, C. C., Palumbo, A. V., Smith, J. C., Wall, J. D., Elias,

D. A. & Liang, L. (2013). The genetic basis for bacterial mercury methylation. *Science*, 339, 1332-1335. https://doi.org/10.1126/science.1230667.

Perez, D. E. L., Ángel, M. L. H., Gomez, M. A. S. & Cartagena, C. J. (2014). Validation of an analytical method for the determination of mercury in shrimp and fish. *Journal of the American Chemical Society*, 109, 1661-1665.

Praveena, S. M., Burbure, C.D., Aris, A. Z. & Hashim, Z. (2013). Mini review of mercury contamination in environment and human with an emphasis on Malaysia: status and needs. *Reviews on Environmental Health* 28: 195-202. https://doi.org/10.1515/reveh-2013-0011.

Rahimi, E., Hajisalehi, M., Kazemeini, H. R., Chakeri, A. & Kashkahi, M. F. (2010). Analysis and determination of mercury, cadmium and lead in canned tuna fish marketed in Iran. *African Journal of Biotechnology*, 9(31), 4938-4941.

Ramezani, Z., Behfar, A. A. & Jafari, S. (2010). Low density polyethylene window for the determination of mercury by cold vapor atomic absorption spectrometry. *Journal of Analytical Atomic Spectrometry*, 25, 886-888. https://doi.org/10.1039/C001080D.

Reyes, L. H., Rahman, G., Fahrenholz, T. & Kingston, H. (2008). Comparison of methods with respect to efficiencies, recoveries, and quantitation of mercury species interconversions in food demonstrated using tuna fish. *Analytical and Bioanalytical Chemistry*, 390(8), 2123-2132. https://doi.org/10.1007/s00216-008-1966-3.

Rice, K. M., Walker Jr., E. M., Wu, M., Gillette, C. & Blough, E. R. 2014. Environmental mercury and its toxic effects. *Journal of Preventive Medicine & Public Health*. 47: 74-83. https://doi.org/10.3961/jpmph.2014.47.2.74.

Romero, V., Costas-Mora, I., Lavilla, I. & Bendicho, C. (2011). Cold vapor-solid phase microextraction using amalgamation in different Pd-based substrates combined with direct thermal desorption in a modified absorption cell for the determination of Hg by atomic absorption spectrometry. *Spectrochimica Acta Part B Atomic Spectroscopy*, 66(2), 156-162. https://doi.org/10.1016/j.sab.2011.01.005.

Saito, H. (2020). Congenital Minamata disease: a description of two cases in Niigata. *NeuroToxicology*, 81, 360-363. https://doi.org/10.1016/j.neuro.2020.09.030.

Saito, H., Sekikawa, T., Taguchi, J., Shozawa, T., Kinoshita, Y. & Matsumura, K. (2020). Prenatal and postnatal methyl mercury exposure in Niigata, Japan: adult outcomes. *NeuroToxicology*, 81: 364-372. https://doi.org/10.1016/j.neuro.2020.09.031.

Strode, S. A., Jaeglé, L., Selin, N. E., Jacob, D. J., Park, R. J., Yantosca, R. M., Mason, R. P. & Slemr, F. (2007). Air-sea exchange in the global mercury cycle. *Global Biogeochemical Cycles* 21: GB1017. https://doi.org/10.1029/2006GB002766, 2007.

UNEP (United Nations Environment Programme). 2019a. Global Mercury Assessment 2018. UN Environment Programme, Chemicals and Health Branch Geneva, Switzerland.

UNEP (United Nations Environment Programme). 2019b. Technical background report to the global mercury assessment 2018. United Nations Environment Programme/ Artic Monitoring and Assessment Programme. https://www.unep.org/resources/publication/global-mercury-assessment-2018.

Ural, M., Uysal, K., Cicek, A., Köse, E., Koçer, M. A. T., Area, S., Örnekçi, G. N. & Yüce, S. (2011). Determination of trace element concentrations in water, sediment and fish species from the Atatürk Dam Lake (Euphrates), Turkey. *Fresenius Environmental Bulletin*, 20(8a), 2036 – 2040.

USEPA. 2007. EPA method 7473 (SW-846): Mercury in solids and solutions by thermal decomposition, amalgamation, and atomic absorption spectrophotometry. Environmental Sampling and Analytical Methods (ESAM) Program. https://www.epa.gov/esam/epa-method-7473-sw-846-mercury-solids-and-solutions-thermal-decomposition-amalgamation-and.

Voegborlo, R. B. & Adimado, A. A. (2010). A simple classical wet digestion technique for the determination of total mercury in fish tissue by cold-vapour atomic absorption spectrometry in a low technology environment. *Food Chemistry*, 123(3), 936-940. https://doi.org/10.1016/j.foodchem.2010.04.059.

Williams, C. R., Leaner, J. J., Somerset, V. S. & Nel, J. M. (2011). Mercury concentrations at a historically mercury-contaminated site in KwaZulu-Natal (South Africa). *Environmental Science and Pollution Research*, 18, 1079-1089. https://doi.org/10.1007/s11356-011-0458-8.

Xiao, J. (2004). Sample preparation and heavy metal determination by atomic spectrometry. Brock University.

Yamashita, Y., Amlund, H., Suzuki, T., Hara, T., Hossain, M. A., Yabu, T., Touhata, K. & Yamashita, M. (2011). Selenoneine, total selenium, and total mercury content in the muscle of fishes. *Fisheries Science*, 77(4), 679-686. https://doi.org/10.1007/s12562-011-0360-9.

Yorifuji, B. 2012. Wonderful life with the elements. Kagaku Dojin Publishing Company. Japan.

Chapter 5

The Bioremediation of Insecticides by Microorganisms

Jasnizat Saidin[1,2,*]
Nur Shafinaz Abu Bakar[2]
Nor Atikah Mohamed Zin[1]
and Noraznawati Ismail[1]
[1]Institute of Climate Adaptation and Marine Biotechnology, Universiti Malaysia Terengganu, Kuala Nerus, Terengganu, Malaysia
[2]Faculty of Science and Marine Environment, Universiti Malaysia Terengganu, Kuala Nerus, Terengganu, Malaysia

Abstract

Millions of tonnes of pesticides are released into the environment each year by agricultural activities. These insecticides will also end up in the aquatic environment due to the natural leaching process. The pollution by these pesticides caused various long term negative impacts to the environment and living organisms. Bioremediation is one of the solutions to cope with the pesticide's pollutant. It is a biological method that used microorganisms to degrade organic wastes under controlled conditions. Bioremediation has become popular and well-adapted as it provides a lot more advantages as the operation cost are low, relatively safe, and environmentally friendly. Various microorganisms across genera were discovered to have the ability to facilitate the bioremediation process as they take up the pollutants as their sole carbon source for their growth. It gives benefit to both parties where our environment can be saved, and the microorganisms will get energy for their needs. This process

---

* Corresponding Author's Email: ijaxzt@umt.edu.my.

In: Climate Change and Environmental Perturbations
Editors: T. Min Pau, M. Abd. Ghaffar, M. Danish-Daniel Abdullah et al.
ISBN: 979-8-89113-271-9

happened as a part of the metabolic process of microorganisms which results in the clean-up of polluted soil and water. Based on previous studies, "chlorpyrifos," "lindane" and "malathion" are the most used insecticides in agricultural industries. As for the microorganisms used for biodegradation, *Bacillus* spp. are the most effective to degrade insecticides with a minimum of 75% biodegradation of malathion in a week. Other microorganisms that show a high ability to degrade insecticides are *Pseudomonas* spp., *Streptomyces* spp. and *Serratia* spp. The literature provides evidence that bioremediation is effective in removing pollutants, hence can be further applied with more advanced technologies.

Keywords: DTT, chlorpyrifos, lindane, malathion, bioremediation

Introduction

Rani et al., (2014) have stated that the world population is expected to increase at a tremendous rate, based on the escalating daily rate of birth. To feed this population with the exhausted land area and depleting water sources is a huge concern and challenge for future generations. As the demand for food sources increases, especially from the agriculture industry, the usage of pesticides will also increase. It is estimated that 4 million tonnes of pesticides are used every year on crops to control pests. This results in a high level of residues on crops, which may bring great harm to consumers (Goda et al., 2010). Pesticides protect crops from damage by rodents, insects, and weeds (Rani and Dhania, 2014). Pesticides are classified based on their target pest, such as rodenticides, algaecides, nematicides, and repellents (Yadav and Devi, 2017), but the most common types that are used in Malaysia—specifically, in the agricultural areas of the Cameron Highland—are fungicides, insecticides, and herbicides (Mispan et al., 2015). However, if they are extensively used, it may result in their accumulation in the agricultural produce (Rani and Dhania, 2014). These pesticides are highly toxic, persistent, and can travel through the air and water for long distances, leading to pollution (Pravin et al., 2012). High doses of pesticides could lead to gastrointestinal problem, dizziness, hypertension and, in severe cases, coma (Sullivan and Blose, 1992; Alvarez et al., 2017). Other side-effects may include skin disorders, miscarriages, and cardiopulmonary arrest. In addition to affecting human health, pesticide residues also affect beneficial organisms such as bees, spiders, and earthworms that contribute towards a healthy ecosystem (Bag, 2000; Rani and Dhania, 2014). Besides, they may be resistant to photolytic, chemical, and biological degradation. Thus, there is a need to degrade them, in order to protect the ecosystem (Pravin et al., 2012). Bioremediation is considered a potential method and solution for the degradation of organophosphate molecules, as it is a viable, safe, and environmentally friendly approach. It is also said to be an attractive alternative to other conventional techniques (Saaty and Booth, 1994; Goda et al., 2010). Furthermore, it is a natural process, thus being widely accepted by the public as a waste treatment method for soil that has been contaminated (Vidali, 2001). In addition, certain bacteria have evolved to tolerate hazardous compounds such as pesticides, particularly in marine environments (Ghodsi et al., 2011). The high amounts of pesticides used for pest control in agricultural fields is a major problem, leading to the pollution of soil and water ecosystems. There is also limited information on the levels of pesticide residues in wastewater environments, especially in developing countries (Ismail et al., 2011). Considering the above, this review is focused on assessing the efficacy of bioremediation processes for pesticide residues using micro-organisms. These bioremediation processes involve naturally occurring micro-organisms that can be used to remove pesticides from the environment (Abdel Razek et al., 2019). Bacteria are used in the bioremediation process, as they are widely recognised for their capability to adapt under selective environmental pressure, given that they can convert chemical pollutants such as pesticides into energy and food.

Bioremediation

Bioremediation is a method of biologically degrading organic wastes under controlled conditions. It can also be defined as the use of naturally occurring bacteria, fungi, algae, and plants to break down pollutants that are harmful to humans and the environment (Murali et al., 2014; Rani and Dhania, 2014). Natural attenuation, biostimulation, and bioaugmentation are some of the naturally occurring mitigating processes included in this technology. The biodegradation of a compound is frequently the result of the actions of several organisms, and bioaugmentation occurs when micro-organisms are imported to a contaminated site to speed up the deterioration of the contaminants (Murali et al., 2014; Rani and Dhania, 2014). Living organisms change contaminant molecules through reactions that occur as part of their metabolic process, resulting in cleaning of the polluted land and water (Rani and Dhania, 2014). Due to its potential to eliminate the hazards of many contaminants, bioremediation as a biotechnological process using micro-organisms has become a popular area in microbiology (Ahmad and Ahmad, 2014). Bioremediation has several advantages, including being a natural process that has been widely acknowledged as a waste treatment technique for polluted materials, as well as being less expensive than conventional technologies (Vidali, 2001). Chemical precipitation, activated sludge, carbon adsorption, and microfiltration are some of the physical and chemical processes used in traditional wastewater treatment (Abdel-Razek et al., 2019). These techniques are used to remove inorganic phosphates, organic wastes, and poisons; however, they are costly and environmentally unfriendly. Bioremediation, on the other hand, offers an alternative environmentally friendly technique for the removal of toxins (Abdel-Razek et al., 2019). This is mainly due to its many advantages over other bioremediation techniques such as phytoremediation, vermiremedation, and entomoremediation; as such, microbial bioremediation takes precedence (Demnerova et al., 2005; Ahmad and Ahmad, 2014). Most micro-organisms involved in bioremediation exhibit reactions through enzymatic pathways as biocatalysts to facilitate the biochemical reactions for pollutant treatment. These micro-organisms act against the pollutants only in the presence of nutrients such as nitrogen and phosphorus (Abatenh et al., 2017).

Insecticides

Insecticides are a form of pesticide that can be divided into several categories: Organochlorines, organophosphates, carbamates, and pyrethroids (Kaur et al., 2019). Most organochlorine pesticides are utilized as insecticides to control a wide range of insects but, unfortunately, their residues can persist in the environment for a long time. DDT (1,1-Dichloro-2,2-bis-(4-chlorophenyl) ethane), lindane, endosulfan, aldrin, dieldrin, and chlordane are some of the most common insecticides in this group. Though most industrialized countries, including the United States, have banned the production and application of DDT, it is still utilized for vector control (especially for malaria), mostly in third-world countries (Yadav and Devi, 2017). Organophosphorus pesticides include parathion, malathion, diazinon, fenitrothion, and chlorpyrifos. These organophosphate pesticides are known as broad-spectrum pesticides, due to their multiple functions in the control of various types of pests. As cholinesterase inhibitors, organophosphorus pesticides are more hazardous to vertebrates and invertebrates alike, resulting in the permanent presence of acetylcholine neurotransmitters in the synapse. As such, nerve impulses are unable to cross the synapse, resulting in rapid contracting action of muscles, paralysis, and death (Yadav and Devi, 2017). Synthetic pyrethroid pesticides are a type of organic pesticide that are synthesized based on the natural pesticide pyrethrin, which is produced by chrysanthemum flowers. Insecticides containing synthetic pyrethroids are extremely poisonous to insects and fish, but only moderately dangerous to mammals and birds. Many synthetic pesticides are non-persistent and readily break down when exposed to sunlight. The synthetic pyrethroid pesticides cypermethrin and permethrin are the most widely used (Yadav and Devi, 2017).

Utilization of Micro-Organisms for Pesticide Bioremediation

A previous study conducted by Abdel-Razek et al., (2019) focused on investigating the potential of microalgae and cyanobacteria, such as *Chlorella vulgaris, Scenedesmus quadricuda*, and *Spirulina platensis*, for the degradation of malathion collected from samples of urban wastewater and agriculture drainage water in Egypt. The results of this study indicated that the fastest algal growth was observed in a treatment consisting of a micro-organismal consortium, malathion, and heavy metals cultured from both water sources. These results corresponded with other studies showing that micro-organisms can utilize various types of organic pollutants which include pesticides as their source of energy to grow, and can simultaneously mineralize and degrade the compounds (Massoud et al., 2008; Abdel-Razek et al., 2019). Besides, the microalgae studied were capable of degrading malathion, nickel, lead, and cadmium from wastewater samples with up to 99%, 95%, 89%, and 88% removal, respectively. The results of this study suggested that a consortium of *Chlorella vulgaris*, *Scenedus quadricuda*, and *Spirulina platensi*s could be used in the bioremediation process of the pesticide malathion and the heavy metals cadmium, nickel and lead, simultaneously and in a very effective manner (Abdel-Razek et al., 2019).

The study of Andleeb et al., (2013) focused on the investigation of malathion degradation by the bacteria *Pseudomonas* spp. Random soil samples were subjected to malathion exposure and isolation processes, resulting in four bacterial isolates. This indicated that the bacteria could adapt to long-term malathion exposure in the toxic man-made environment. This study proved that microbes could successfully utilize a pollutant for its survival requirement and that the isolates (*Pseudomonas* spp.) were good candidates for degrading malathion- or other pesticide-contaminated soils (Andleeb et al., 2013). *Pseudomonas* spp. have been reported to have 100% efficiency for malathion degradation (Rosenberg and Alexander 1979; Andleeb et al., 2013). However, overall, *Bacillus* spp. is the major genus (19%) used in bioremediation studies, followed by *Streptomyces* and *Pseudomonas* (14%); see Figure 1.

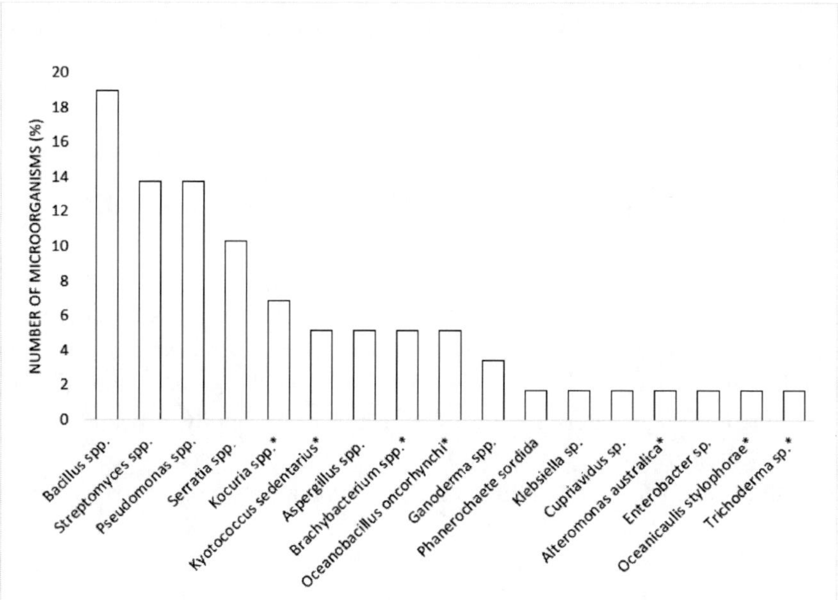

*denotes micro-organisms isolated from a marine environment.

Figure 1. Comparative frequency (%) of Bacterial genera used in bioremediation studies.

Types of Insecticides Used in Biodegradation Studies

An assessment of studies from 2000 to 2020 indicated that many types of insecticides are produced by the industry, with some considered in biodegradation studies. These insecticides are grouped into several categories, including organophosphorus, organochlorine, and pyrethroid insecticides (Figure 2).

Chlorpyrifos was the most widely considered insecticide, making up 29.5% of the total, followed by lindane (16%) and malathion (13.6%).

Chlorpyrifos has been used since the 1960s for house and garden usage (Worthing, 1979; Chishti et al., 2012). It is a non-systemic insecticide and helps in controlling various pests on vegetable crops (Fang et al., 2006; Chishti et al., 2012). Lindane has been used worldwide for fruits, vegetables, crops, animals, and in animal premises over the last 70 years (Singh and Singh, 2019a; Zhang et al., 2020). It can persist in the environment, as it is a lipophilic substance with high chemical stability that can spread over long distances (Benimeli et al., 2008). Malathion is commonly used in agriculture to eradicate sucking and chewing insects on crops and livestock (Latifi et al., 2012). Other insecticides, such as Pyrethroid, α-Endosulfan, diazinon, parathion, and prefenofos, can also be degraded by micro-organisms as described in the literature. Insecticides categorized as "Others" in Figure 2 include chlotianidin, fenithrotion, fenpropathrin, methoxychlor, chlordane, fluroxypyr, DDD (1,1-Dichloro-2,2-bis-(4-chlorophenyl) ethane), and dieldrin, which are less common and rarely considered in bioremediation research.

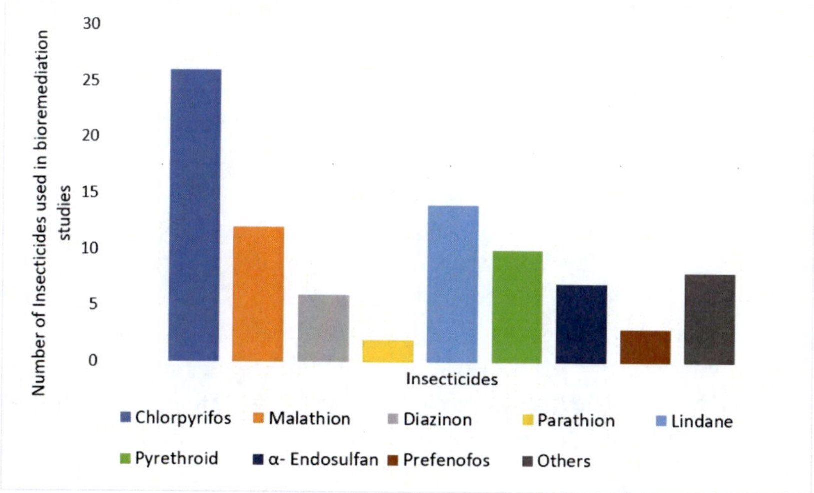

Figure 2. Types of insecticides used in insecticide bioremediation studies.

Source of Micro-Organisms Used in Bioremediation Studies

Most of the micro-organisms (68%) used in the bioremediation studies are terrestrial, while the rest are obtained from marine environments (32%)

(Figure 3). Soil is the major source of micro-organisms at 41%, followed by sponges and coral at 17% and 11%, respectively (Figure 4). Other sources of micro-organisms include sludge, plants, wastewater, paddy fields, rivers, and wells for terrestrial and seawater and sargassum for marine micro-organisms.

Based on Figure 4, soil is the major source of micro-organisms used in bioremediation studies. This is supported by the observations of various kinds of micro-organisms containing organophosphate-degrading enzymes found in soils that have previously been exposed to pesticides (Sasikala et al., 2012; Derbalah et al., 2013). Insecticides containing organophosphorus can either be applied directly to the soil to control pests, or can be introduced indirectly by spray drift or precipitation from the atmosphere. Organophosphorus insecticides could also impact soil quality through a reduction of biodiversity in the soil, as small non-targeted insects may also be killed in addition to the intended pest (Derbalah et al., 2013). Micro-organisms extracted from soil are used for bioremediation processes as they are well-known for having the capability to adapt to environments under selective pressure, and can utilise these toxic chemicals as a source of energy (Andleeb et al., 2013). Enzymes derived from marine micro-organisms have more advantages, when compared to those originating from plants or animals. They have a wide range of catalytic activities, are solvent- and temperature-stable, generate a lot of products in a short amount of time, and are simple to harvest (Sivaperumal et al., 2017). Some archaea, fungi, and bacteria found in the ocean could accelerate the bioremediation process. Furthermore, enzyme-assisted methods are quickly gaining popularity due to the shorter time required to run the process, lower energy requirements, less hazardous, cost-effective, and environmentally friendly characteristics (Yang et al., 2012; Sivaperumal et al., 2017). As stated by Aresta et al., (2015), bacterial strains isolated from the sponge *Hymeniacidon perlevis* can break down lindane, and it is believed to have already established a symbiotic relationship with the sponge. It is also a well-known fact that the ability of microbes to breakdown lindane *in situ* can be improved by the process of induction and acclimation, in which the enzyme systems of biodegradation pathways are increased, allowing for more efficient pollutant removal (Elcey and Kunhi, 2010; Aresta et al., 2015). The rate of contaminant degradation is directly related to the concentration of contaminants and the amount of catalysts present in the environment. These catalysts refer to the number of viable organisms that can metabolize the contaminants in the presence of enzyme, thus resulting in an increase or decrease in the contaminant degradation rate (Abatenh et al., 2017). As mentioned before, bioremediation can only effectively progress when the environmental conditions are favourable for microbial growth in the presence of the catalyst. Its application as a pollutant treatment tool often involves the manipulation of several environmental parameters to permit microbial growth and degradation at a faster rate (Kumar et al., 2011). At present, researchers tend to opt for biotechnological and molecular approaches to increase the bioremediation rate through genetic engineering. The involvement of genetically engineered micro-organisms (GEM) has improved the utilization and elimination of hazardous pollutants in most environmental conditions (Jain et al., 2010).

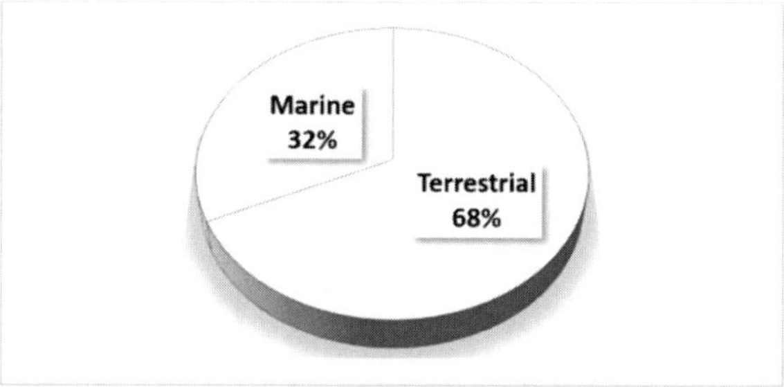

Figure 3. Sources of micro-organisms used in insecticide bioremediation processes.

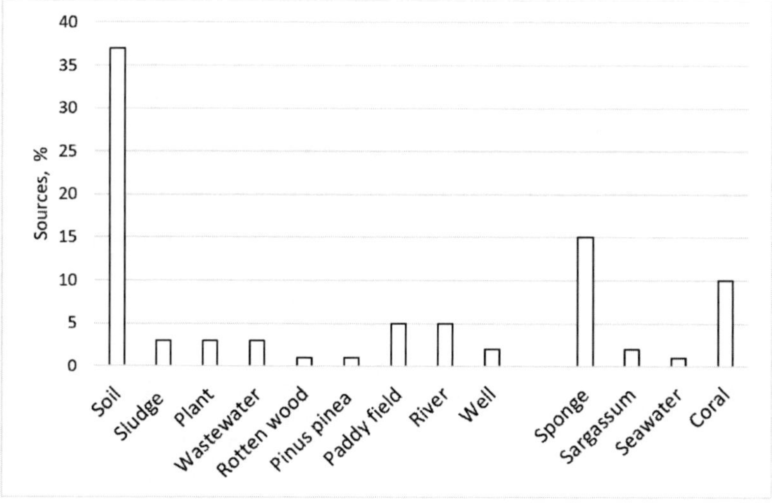

Figure 4. Sources of micro-organisms used in insecticide bioremediation processes.

Table 1. Chlorpyrifos degradation by various micro-organisms

Micro-organism	Source	Insecticides	Results	References
Pseudomonas aeruginosa	Soil	Chlorpyrifos	10, 25, 50 mg/l: 100% degraded, 75 mg/l: 52% degraded, 100 mg/l: 25% degraded	(Fulekar & Geetha, 2008)
Cupriavidus sp.	Sludge	Chlorpyrifos and 3,5,6-trichloro-2-pyridinol	100% degradation of chlorpyrifos and 94.3% degradation of TCP (trichloro pyridinol)	(Lu et al., 2013)
Serratia marcescens	Soil	Chlorpyrifos	58.9% degradation of 50 mg/l chlorpyrifos within 14 days	(Cycon et al., 2013)
Pseudomonas sp.	Soil	Chlorpyrifos	91% degradation of 300 mg L^{-1} d^{-1} in 7 days	(Yadav et al., 2014)
Ganoderma sp. JAS4	Soil	Chlorpyrifos	Degradation rate constant of 0.8460 day^{-1} (aqueous medium)	(Silambarasan & Abraham, 2012)
Ochrobactrum sp. JAS2	Soil	Chlorpyrifos	100% degradation of 300 mg/L chlorpyrifos after 70 h	(Silambarasan & Abraham, 2016)
Aspergillus terreus JAS1	Soil	Chlorpyrifos	Degraded chlorpyrifos and its metabolite TCP (300 $mgkg^{-1}$ soil) in 24 and 48 h	(Silambarasan & Abraham, 2013)
Synechocystis sp. strain PUPCCC 64	Soil	Chlorpyrifos	3.78 mg/L and 4.69 mg/L of chlorpyrifos removed in 4 and 8 days	(Singh et al., 2011)
Bacillus pumilus C2A1	Soil	Chlorpyrifos	89% degradation of 1000 mg l^{-1} chlorpyrifos degraded within 15 days	(Anwar et al., 2009)
Klebsiella sp	Wastewater	Chlorpyrifos	92% degradation of 3.84 g l^{-1} chlorpyrifos in 4 days	(Ghanem et al., 2007)
Paracoccus sp. strain TRP	Sludge	Chlorpyrifos	100% degradation of 50 mg l^{-1} chlorpyrifos in 4 days	(Xu et al., 2008)
Pseudomonas nitroreducens PS-2	Soil	Chlorpyrifos	100% degradation of 100 mg l^{-1} chlorpyrifos after 28 days	(Korade et al., 2009)
Serratia sp.	Sludge	Chlorpyrifos	100% degradation of 50 mg l^{-1} chlorpyrifos culture by a co-culture in 18 h	(Xu et al., 2007)
Sphingomonas sp.	Soil & water	Chlorpyrifos	100% degradation of 100 mg l^{-1} chlorpyrifos in 2 days (liquid culture) and 98% degradation of 30 mg L^{-1} in 10 days (soil)	(Li et al., 2007)
Enterobacter sp.	Soil	Chlorpyrifos	100% degradation of 100 mg l^{-1} in 10 days in liquid culture	(Singh et al., 2003)

Micro-organism	Source	Insecticides	Results	References
Gordonia sp. *JAAS1*	Paddy field	Chlorpyrifos	100% degradation of 110 mg/L chlorpyrifos in 24 h	(Abraham et al., 2013)
Brachybacterium sp. K473	Coral	Chlorpyrifos	22.4% degradation of 20 μg/L chlorpyrifos in 5 days	(Radjasa & Sabdono, 2008)
Kytococcus sedentarius 13	Coral	Chlorpyrifos	12.22% degradation of 20 μg/L chlorpyrifos in 5 days	
Oceanobacillus oncorhynci R-2	Coral	Chlorpyrifos	44.4% degradation of 20 μg/L chlorpyrifos in 5 days	
Pseudomonas aeruginosa	Paddy field	Chlorpyrifos	92% degradation of 50 mg/L chlorpyrifos in 20 days in liquid medium	(Lakshmi et al., 2008)
Pseudomonas fluorescence	Paddy field	Chlorpyrifos	89% degradation of 50 mg/L chlorpyrifos in 20 days in liquid medium	
Brucella melitensis	Paddy field	Chlorpyrifos	87% degradation of 50 mg/L chlorpyrifos in 20 days in liquid medium	
Bacillus subtilis	Paddy field	Chlorpyrifos	85% degradation of 50 mg/L chlorpyrifos in 20 days in liquid medium	
Bacillus paramycoides LPA2.1	Well	Chlorpyrifos	28.5% degradation of 50 mg /L chlorpyrifos in 4 days	
Acremonium sp. (GFRC-1)	Soil	Chlorpyrifos	84% degradation of 300 mg/L chlorpyrifos in 20 days	(Kulshrestha & Kumari, 2011)

Table 2. Lindane degradation by various micro-organisms

Micro-organism	Source	Insecticides	Results	References
Ganoderma australe	*Pinus pinea*	Lindane	Biodegradation/biomass of 56.9 µg/g	(Rigas et al., 2007)
Mameliella phaeodactyli	*Hymeniacidon perlevis* (Sponge)	Lindane	89% degradation of 0.05 mg/L lindane in 12 days	(Loredana et al., 2017)
Pseudovibrio ascidiaceicola		Lindane	95% degradation of 0.05 mg/L lindane in 12 days	
Oceanicaulis stylophorae		Lindane	90% degradation of 0.05 mg/L lindane in 12 days	
Alteromonas australica		Lindane	97% degradation of 0.05 mg/L lindane in 12 days	
Colony Li-6	*Hymeniacidon perlevis* (demosponge)	Lindane	97% degradation of 1 lg/L lindane in 8 days	(Aresta et al., 2015)
Colony Li-8		Lindane	95% degradation of 1 lg/L lindane in 8 days	
Colony Li-13		Lindane	97% degradation of 1 lg/L lindane in 8 days	
Streptomyces sp	Soil	Lindane	40.4% degradation of 1.66 mg/L lindane in 16 days in liquid medium	(Fuentes et al., 2017)
Streptomyces A5-M7	Soil	Lindane	32.6% degradation of 1.66 mg/L lindane in 28 days	(Fuentes et al., 2011)
Streptomyces A2-A5-A8	Soil	Lindane	33.1% degradation of 1.66 mg/L lindane in 28 days	
Streptomyces A2-A5-A11-M7	Soil	Lindane	31.4% degradation of 1.66 mg/L lindane in 28 days	
Streptomyces sp. M7	Wastewater	Lindane	17% degradation of 100 µg/L of lindane in 48 h	(Benimeli et al., 2006)
Streptomyces sp. M7	Soil	Lindane	56% degradation of 100 µg/kg lindane in 4 weeks	(Benimeli et al., 2008)

Table 3. Malathion degradation by various micro-organisms

Micro-organism	Source	Insecticides	Results	References
Bacillus sp.	Soil	Malathion	Sterile 84.81%, non-sterile 74.11% degraded from total of 1.5% kg^{-1} soil after 7 days	(Singh et al., 2013)
Pseudomonas spp	Cucumber leaves	Malathion	0.10 ± 0.001 μg/ml d^{-1}	(Massoud et al., 2007)
Asperigillus spp.	Cucumber leaves	Malathion	0.12 ± 0.01 μg/ml d^{-1}	
Penicilluim spp.	Cucumber leaves	Malathion	0.11 ± 0.02 μg/ml d^{-1}	
Bacillus licheniformis	Soil	Malathion	78% of 350 μg/mL degraded within 5 days	(Khan et al., 2016)
Lysinibacillus sp	Soil	Malathion	20% malathion and 47% malaoxon degraded after 7 days	(Singh et al., 2012)
Bacillus sp. S14	Soil	Malathion	68% of 50 mg/l degraded after 20 h	(Adhikari et al., 2011)
Brevibacillus sp. strain KB2	Soil	Malathion	36.22% of malathion degraded after 7 days	(Singh et al., 2012)
Bacillus cereus strain PU	Soil	Malathion	87.40% of malaoxon and 49.31% of malathion degraded after 7 days	(Singh et al., 2012)
Bacillus spp.	Soil	Malathion	100% degradation after 48 h	(Kumari et al., 2012)
Pleurotus ostreatus	Soil	Malathion	76.4% of 50 μg/100 mL degraded after 25 days	(Ganash et al., 2016)
Pseudomonas putida	Soil	Malathion	100% of 50 mg/l degraded after 6 days	(Goda et al., 2010)

Insecticide-Degrading Micro-Organism Species Isolated from Marine and Terrestrial Environments

Bioremediation using micro-organisms is a common process for the removal of organic pollutants, as it is less expensive, safer, and environmentally friendly (Saaty and Booth, 1994; Goda et al., 2010). Many species of micro-organisms have been reported to degrade insecticides. Table 1 to Table 3 detail the degradation rates of chlorpyrifos, lindane, and malathion, respectively, which are also the most-used insecticides in insecticide bioremediation studies. Various species of micro-organisms have been found to degrade these insecticides successfully.

Chlorpyrifos-Degrading Micro-Organisms Isolated from Marine and Terrestrial Environments

Biodegradation of chlorpyrifos can occur both in liquid medium and soil. Water has been found to contain 1,2,3-Trichloropropane (TCP), a key metabolic by-product of chlorpyrifos which has a longer half-life than chlorpyrifos. It is also one of the limiting variables for the mineralization of chlorpyrifos in aquatic environments. Bacterial species that may degrade chlorpyrifos in liquid medium come from various genera, such as *Enterobacter, Alcaligenes, Klebsiella, Providencia, Pseudomonas, Leuconostoc, Lactobacillus, Serratia*, and *Synechocysti* (Singh et al., 2011; Chishti et al., 2013). Bacteria and fungi can break down chlorpyrifos by utilising it as a direct source of carbon, nitrogen, and phosphorus, or by co-metabolizing it (Chishti et al., 2013). A micro-organism transforms a target compound without the need to make use of the constituent elements as the sole carbon source or energy in co-metabolism. In addition, microbial species that have the capability to utilize chlorpyrifos can be extracted using enrichment culture techniques from chlorpyrifos-contaminated soil, sludge, wastewater, paddy fields, and wells. *Pseudomonas* spp. are the most frequently used, as they have a high degradation efficiency (ranging from 89–100%). A wide range of micro-organisms can decompose chlorpyrifos. Pesticide-degrading enzymes such as organophosphorus hydrolase (OPH), phosphotriesterase (PTE), acid organophosphorus anhydrolase (OPAA), and methyl parathion hydrolase (MPH) are produced by these bacteria (Chishti et al., 2013). Chlorpyrifos breaks down into two major metabolites—namely, TCP and

DETP—depending on the condition of the micro-organisms and their environment, while other metabolites—including desethyl chlorpyrifos, chlorpyrifosoxon, desethyl chlorpyrifos-oxon, and 3,5,6-trichloro-2 methoxypyrimidine—are produced in minute amounts. In the soil, TCP is thought to be both mobile and permanent (Chishti et al., 2013). In a liquid medium, *Pseudomonas* spp. were found to mineralize TCP through the reductive de-chlorination route (Feng et al., 1997; Chishti et al., 2013).

Lindane-Degrading Micro-organisms Isolated from Marine and Terrestrial Environments

A list of lindane-degrading micro-organism is provided in Table 2, including *Ganoderma australe* (Rigas et al., 2007), *Mameliella phaeodactyli, Pseudovibrio ascidiaceicola, Oceanicaulis stylophorae, Alteromonas australica* (Loredana et al., 2017), Colony Li-6, 7, 8, 13 (Aresta et al., 2015), and *Streptomyces* spp. (Benimeli et al., 2006; Benimeli et al., 2008; Fuentes et al., 2011; Fuentes et al., 2017). These micro-organisms play a vital role in the bioremediation process, through either chemical or physical interactions that lead to structural alterations of or fully breaking down the target molecules. According to reports, different micro-organisms break down lindane at varying rates. Based on Table 2, *Streptomyces* sp. M7, which was isolated from wastewater, had the lowest degradation rate efficiency (58% of 100 μg/L lindane over a four-week period). Meanwhile, micro-organisms isolated from the sponge *Hymeniacidon perlevis* showed a very high degradation rate efficiency ranging from 89% to 97%. As stated by Sivaperumal et al., (2017), marine micro-organisms have more advantages, as they present a wide range of catalytic activity, solvent- and temperature-stability, and can produce many products in a short amount of time. Microbial metabolism can break down lindane effectively by creating extracellular and intracellular enzymes that convert lindane into less-toxic compounds. It has been reported that micro-organisms can degrade lindane under both aerobic and anaerobic conditions; however, it is mostly degraded only under aerobic conditions (Alvarez et al., 2012; Zhang et al., 2020). The most significant step for the degradation of lindane by micro-organisms is the removal of the chlorine atom. In this pathway, lindane will be converted to 1,2,4-trichlorobenzene (1,2,4-TCB), 2,5-dichlorophenol (2,5-DCP), and 2,5-dichlorohydroquinone (2,5-DCHQ) by the enzymatic activities of dehydrochlorinase (LinA), halidohydrolase (LinB), and dehydrogenase (LinC). The reduction of lindane to 2,5-DCHQ is called the upstream process, which is then metabolized by the downstream pathway (Endo et al., 2005; Zhang et al., 2020).

Malathion-Degrading Micro-organisms Isolated from Terrestrial Environments

Malathion was the earliest high-selectivity toxic organophosphorus insecticide used in residential areas and for agriculture (Singh et al., 2012). To date, several malathion-degrading micro-organisms have been screened and analysed (Table 3) including *Bacillus* spp. (Kumari et al., 2012; Singh et al., 2013), *Pseudomonas* spp., *Aspergillus* sp., *Penicillium* sp. (Massoud et al., 2007), *Bacillus licheniformis* (Khan et al., 2016), *Lysinibacillu*s sp, *Brevibacillus* sp. strain KB2, *Bacillus cereus* strain PU (Singh et al., 2012), *Bacillus* sp. S14 (Adhikari et al., 2011), *Pleurotus ostreatus* (Ganash et al., 2016), and *Pseudomonas putida* (Goda et al., 2010). These micro-organisms can use malathion as a sole carbon source (Singh et al., 2012). The research conducted by Singh et al., (2013) showed that malathion can degrade better in inoculated sterilized soil, reaches 70% degradation compared to 7.11% in non-inoculated sterilized soil. Massoud et al., (2007) extracted the micro-organisms *Bacillus licheniformis, Aspergillus* sp., and *Penicillium* sp. from cucumber leaves. The presence of hydrolytic enzymes (esterases) may be the main reason for malathion degradation in this case. They also found that the growth response of malathion by *Aspergillus* sp. and *Penicillium* sp. increased proportionately with malathion degradation (Massoud et al., 2007). This reveals the key step in overcoming the malathion residues in vegetable crops, as these micro-organisms were isolated from cucumber leaves. *Bacillus* spp. are the major micro-organisms used to biodegrade malathion, with success rate ranging from 68% to 87%. This biodegradation process was assisted by the action of carboxylesterase (Singh et al., 2012).

Conclusion

The increasing demand for pesticides has resulted in their increased usage. Bioremediation is a very effective way to degrade organic pollutants such as insecticides. A literature analysis of 50 publications revealed that many micro-organisms have the potential to degrade insecticides. For instance, *Bacillus* spp., *Pseudomonas* spp., *Streptomyces* spp., and *Serratia* spp. are widely known to degrade insecticides. Many researchers have used these micro-organisms as they have demonstrated high efficiency in degrading various types of insecticides. Most of the sourced micro-organisms were of terrestrial origin (covering 68%), while only 32% of the studies indicated that marine micro-organisms can degrade insecticides. Thus, there is still a large gap in the literature related to marine micro-organisms, when compared to those of terrestrial origin. Many different sources of micro-organisms for bioremediation processes are yet to be discovered; for example, from marine and mangrove ecosystems. Moreover, the development of novel technologies globally is expected to provide better tools for the bioremediation of harmful pollutants.

References

Abatenh, E., Gizaw, B., Tsegaye Z. & Wassie, M. (2017). The Role of Microorganisms in Bioremediation- A Review. *Open Jorunal of Environmental Biology*, 2(1): 038-046.

Abdel-Razek, M. A., Abozeid, A. M., Eltholth, M. M., Abouelenien, F. A., El-Midany, S. A., Moustafa, N. Y. & Mohamed. R. A. (2019). Bioremediation of a pesticide and selected heavy metals in wastewater from various sources using a consortium of microalgae and cyanobacteria. *Slovenian Veterinary Research*, 56(22), 61-73.

Abo-Amer, A. (2011) Biodegradation of diazinon by *Serratia marcescens* DI101 and its use in bioremediation of contaminated environment. *Journal of Microbiology and Biotechnology*, 71-80.

Abraham, J. & Silambarasan, S. (2016). Biodegradation of chlorpyrifos and its hydrolysis product 3, 5, 6-trichloro-2-pyridinol using a novel bacterium *Ochrobactrum* sp. JAS2: a proposal of its metabolic pathway. *Pesticide Biochemistry and Physiology*, 126, 13-21.

Abraham, J., Shanker, A. & Silambarasan, S. (2013). Role of *Gordonia* sp JAAS 1 in biodegradation of chlorpyrifos and its hydrolysing metabolite 3, 5, 6-trichloro-2-pyridinol. *Letters in Applied Microbiology*, 57(6), 510-516.

Adhikari, S., Chattopadhyay, P. & Ray, L. (2011). Bioremediation of malathion from environment by an isolated *Bacillus* sp. S14. *Indian Chemical Engineer*, 53(2), 68-83.

Ahmad, M. & Ahmad, I. (2014). Bioremediation of pesticides. *Biotechnology*, 11, 125-165.

Andleeb, S., Ghauri, M. A. & Qazi, J. I. (2013). Isolation of malathion degrading chromogenic *Pseudomonas aeruginosa* strains from insecticide impregnated soils. *Pakistan Journal of Zoology*, 45(5), 1447-1451.

Anwar, S., Liaquat, F., Khan, Q. M., Khalid, Z. M. & Iqbal, S. (2009). Biodegradation of chlorpyrifos and its hydrolysis product 3, 5, 6-trichloro-2-pyridinol by *Bacillus pumilus* strain C2A1. *Journal of Hazardous Materials,* 168(1), 400-405.

Aresta, A., Marzano, C. N., Lopane, C., Corriero, G., Longo, C., Zambonin, C. & Stabili, L. (2015). Analytical investigations on the lindane bioremediation capability of the demosponge *Hymeniacidon perlevis*. *Marine Pollution Bulletin*, 90(1-2), 143-149.

Benimeli, C. S., Castro, G. R., Chaile, A. P. & Amoroso, M. J. (2006). Lindane removal induction by *Streptomyces* sp. M7. *Journal of Basic Microbiology*, 46(5), 348-357.

Benimeli, C. S., Fuentes, M. S., Abate, C. M. & Amoroso, M. J. (2008). Bioremediation of lindane-contaminated soil by *Streptomyces* sp. M7 and its effects on Zea mays growth. *International Biodeterioration & Biodegradation*, 61(3), 233-239.

Birolli, W. G., Borges, E. M., Nitschke, M., Romão, L. P. & Porto, A. L. (2016). Biodegradation pathway of the pyrethroid pesticide esfenvalerate by bacteria from different biomes. *Water, Air, & Soil Pollution*, 227(8), 1-11.

Birolli, W. G., Vacondio, B., Alvarenga, N., Seleghim, M. H. & Porto, A. L. (2018). Enantioselective biodegradation of the pyrethroid (±)-lambda-cyhalothrin by marine-derived fungi. *Chemosphere*, 197, 651-660.

Birolli, W. G., Yamamoto, K. Y., de Oliveira, J. R., Nitschke, M., Seleghim, M. H. & Porto, A. L. (2015). Biotransformation of dieldrin by the marine fungus *Penicillium miczynskii* CBMAI 930. *Biocatalysis and Agricultural Biotechnology*, 4(1), 39-43.

Chen, S., Chang, C., Deng, Y., An, S., Dong, Y. H., Zhou, J. & Zhang, L. H. (2014). Fenpropathrin biodegradation pathway in *Bacillus* sp. DG-02 and its potential for bioremediation of pyrethroid-contaminated soils. *Journal of Agricultural and Food Chemistry*, 62(10), 2147-2157.

Chishti, Z., Hussain, S., Arshad, K. R., Khalid, A. & Arshad, M. (2013). Microbial degradation of chlorpyrifos in liquid media and soil. *Journal of Environmental Management, 114*, 372-380.

Cycoń, M., Wójcik, M. & Piotrowska-Seget, Z. (2009). Biodegradation of the organophosphorus insecticide diazinon by *Serratia* sp. and *Pseudomonas* sp. and their use in bioremediation of contaminated soil. *Chemosphere*, 76(4), 494-501.

Cycoń, M., Żmijowska, A., Wójcik, M. & Piotrowska-Seget, Z. (2013). Biodegradation and bioremediation potential of diazinon-degrading *Serratia marcescens* to remove other organophosphorus pesticides from soils. *Journal of Environmental Management*, 117, 7-16.

Derbalah, A. S., Ismail, A. & Hamza, A. (2013). Monitoring of organophosphorus pesticides and bioremediation chlorpyrifos in soil. *Egyptian Journal of Plant Production Research*, 1, 115-132.

Fuentes, M. S., Raimondo, E. E., Amoroso, M. J. & Benimeli, C. S. (2017). Removal of a mixture of pesticides by a *Streptomyces consortium*: Influence of different soil systems. *Chemosphere*, 173, 359-367.

Fuentes, M. S., Sáez, J. M., Benimeli, C. S. & Amoroso, M. J. (2011). Lindane biodegradation by defined consortia of indigenous *Streptomyces* strains. *Water, Air, & Soil Pollution*, 222(1), 217-231.

Fulekar, M. H. & Geetha, M. (2008). Bioremediation of chlorpyrifos by *Pseudomonas aeruginosa* using scale up technique. *Journal of Applied Biosciences*, 12, 657-60.

Ganash, M. A., Abdel Ghany, T. M. & Reyad, A. M. (2016). *Pleurotus ostreatus* as a biodegradator for organophosphorus insecticide—malathion. *Journal of Environmental and Analytical Toxicolology*, 6(369), 2161-0525.

Ghanem, I., Orfi, M. & Shamma, M. (2007). Biodegradation of chlorphyrifos by *Klebsiella* sp. isolated from an activated sludge sample of waste water treatment plant in Damascus. *Folia Microbiologica*, 52(4), 423-427.

Goda, S. K., Elsayed, I. E., Khodair, T. A., El-Sayed, W. & Mohamed, M. E. (2010). Screening for and isolation and identification of malathion-degrading bacteria: cloning and sequencing a gene that potentially encodes the malathion-degrading enzyme, carboxylestrase in soil bacteria. *Biodegradation*, 21(6), 903-913.

Ismail, B. S., Sameni, M. & Halimah, M. (2011). Evaluation of herbicide pollution in the kerian ricefields of Perak, Malaysia. *World Applied Sciences Journal*, 15(1), 05-13.

Jain, P. K., Gupta, V., Gaur, R. K., Bajpai, V., Gautama, N., Modi, D. R. (2010c). Fungal Enzymes: Potential Tools of Environmental Processes. In: Fungal Biochemistry and Biotechnology, Gupta, VK, Tuohy M and Gaur RK (Eds.). *LAP Lambert Academic Publishing AG and Co. KG*, Germany. 44-56.

Jayasree, V. S., Sobhana, K. S., Poulose, P., Babu, K., Jasmine, S., Ranjith, L., Saravanan, R., Kingsly, H. J., Sreentah, K. R., Joshi, K. K., George, R. M. & Gopalakrishnan, A. (2020). Biodegradation of the pyrethroid pesticide cyfluthrin by the halophilic bacterium *Photobacterium ganghwense* isolated from coral reef ecosystem. *Indian Journal of Fisheries*, 67(4), 116-131.

Kafilzadeh, F., Ebrahimnezhad, M. & Tahery, Y. (2015). Isolation and identification of endosulfan-degrading bacteria and evaluation of their bioremediation in Kor River, Iran. *Osong Public Health and Research Perspectives*, 6(1), 39-46.

Kaur, R., Mavi, G. K., Raghav, S. & Khan, I. (2019). Pesticide classification and its impact on environment. *International Journal of Current Microbiology Applied Sciences*, 8, 1889-1897.

Khan, S., Zaffar, H., Irshad, U., Ahmad, R., Khan, A. R., Shah, M. M. & Naqvi, T. (2016). Biodegradation of malathion by Bacillus licheniformis strain ML-1. *Archives of Biological Sciences*, 68(1), 51-59.

Korade, D. L. & Fulekar, M. H. (2009). Rhizosphere remediation of chlorpyrifos in mycorrhizospheric soil using ryegrass. *Journal of Hazardous Materials*, 172(2-3), 1344-1350.

Kulshrestha, G. & Kumari, A. (2011). Fungal degradation of chlorpyrifos by *Acremonium sp.* strain (GFRC-1) isolated from a laboratory-enriched red agricultural soil. *Biology and Fertility of Soils*, 47(2), 219-225.

Kumar, A., Bisht, B. S., Joshi, V. D. & Dhewa, T. (2011). Review on Bioremediation of Polluted Environment: A Management Tool. *International Journal of Environmental Sciences*, 1: 1079-1093.

Kumari, A. R., Jeevan, G., Ashok, M., Rao, C. & Vamsi, K. S. (2012). Malathion degradation by *Bacillus* spp. isolated from soil. *IOSR Journal of Pharmacy*, 2, 37-42.

Lakshmi, C. V., Kumar, M. & Khanna, S. (2008). Biotransformation of chlorpyrifos and bioremediation of contaminated soil. *International Biodeterioration & Biodegradation*, 62(2), 204-209.

Li, X., He, J. & Li, S. (2007). Isolation of a chlorpyrifos-degrading bacterium, *Sphingomonas* sp. strain Dsp-2, and cloning of the mpd gene. *Research in Microbiology*, 158(2), 143-149.

Loredana, S., Graziano, P., Antonio, M., Carlotta, N. M., Caterina, L., Maria, A., Carlo, Z., Giuseppe, C. & Pietro, A. (2017). Lindane bioremediation capability of bacteria associated with the demosponge *Hymeniacidon perlevis*. *Marine Drugs*, 15(4), 108.

Lu, P., Li, Q., Liu, H., Feng, Z., Yan, X., Hong, Q. & Li, S. (2013). Biodegradation of chlorpyrifos and 3, 5, 6-trichloro-2-pyridinol by *Cupriavidus* sp. DT-1. *Bioresource Technology*, 127, 337-342.

Massoud, A. H., Derbalah, A. S., Belal, B. & El-Fakharany, I. I. (2007). Bioremediation of Malathion in aquatic system by different microbial isolates. Journal of *Pest Control Environmental Sciences*, 15, 13-28.

Mispan, M. R., Haron, S. H., Ismail, B. S., Abd Rahman, N. F., Khalid, K. & Rasid, M. A. (2015). The use of pesticides in agriculture area, Cameron Highlands. *Age (years)*, 30(10.3), 22-60.

Mori, T., Wang, J., Tanaka, Y., Nagai, K., Kawagishi, H. & Hirai, H. (2017). Bioremediation of the neonicotinoid insecticide clothianidin by the white-rot fungus *Phanerochaete sordida. Journal of Hazardous Materials*, 321, 586-590.

Ortega, S. N., Nitschke, M., Mouad, A. M., Landgraf, M. D., Rezende, M. O. O., Seleghim, M. H. R., Sette, L. D. & Porto, A. L. M. (2011). Isolation of Brazilian marine fungi capable of growing on DDD pesticide. *Biodegradation*, 22(1), 43-50.

Pailan, S., Gupta, D., Apte, S., Krishnamurthi, S. & Saha, P. (2015). Degradation of organophosphate insecticide by a novel *Bacillus aryabhattai* strain SanPS1, isolated from soil of agricultural field in Burdwan, West Bengal, India. *International Biodeterioration & Biodegradation*, 103, 191-195.

Pravin, D., Sandip, B., Shreyash, B. & Anjana, G. (2012). Degradation of organophosphate and organochlorine pesticides in liquid culture by marine isolate *Nocardiopsis* species and its bioprospectives. *Journal of Environmental Research and Development*, 7(2), 995-1001.

Rani, K. & Dhania, G. (2014). Bioremediation and biodegradation of pesticide from contaminated soil and water – a noval approach. *International Journal of Current Microbiology and Applied Science*, 3(10), 23-33.

Rigas, F., Papadopoulou, K., Dritsa, V. & Doulia, D. (2007). Bioremediation of a soil contaminated by lindane utilizing the fungus *Ganoderma australe* via response surface methodology. *Journal of Hazardous Materials*, 140(1-2), 325-332.

Rochaddi, B., Zainuri, M. & Sabdono, A. (2019). Diversity of chlorpyrifos-degrading bacteria isolated from shallow aquifer of East Java Coastal Settlements, Indonesia. *Biodiversitas Journal of Biological Diversity*, 20(12).

Sabdono, A. & Radjasa, O. K. (2008). Phylogenetic diversity of organophosphorous pesticide-degrading coral bacteria from mid-west coast of Indonesia. *Biotechnology*, 7(4), 694-701.

Sethunathan, N., Megharaj, M., Chen, Z. L., Williams, B. D., Lewis, G. & Naidu, R. (2004). Algal degradation of a known endocrine disrupting insecticide, α-endosulfan, and its metabolite, endosulfan sulfate, in liquid medium and soil. *Journal of Agricultural and Food Chemistry*, 52(10), 3030-3035.

Silambarasan, S. & Abraham, J. (2014). Efficacy of Ganoderma sp. JAS4 in bioremediation of chlorpyrifos and its hydrolyzing metabolite TCP from agricultural soil. *Journal of Basic Microbiology*, 54(1), 44-55.

Silambarasan, S. & Abraham, J. (2013). Ecofriendly Method for Bioremediation of Chlorpyrifos from Agricultural Soil by Novel Fungus *Aspergillus terreus* JAS1. *Water Air Soil Pollution*, 224, 1369. https://doi.org/10.1007/s11270-012-1369-0.

Singh, B, Kaur, J. & Singh, K. (2013) Bioremediation of malathion in soil by mixed *Bacillus* culture. *Advances in Bioscience and Biotechnology*, 4, 674-678. doi: 10.4236/abb.2013.45088.

Singh, B. K., Walker, A., Morgan, J. A. W. & Wright, D. J. (2003). Effects of soil pH on the biodegradation of chlorpyrifos and isolation of a chlorpyrifos-degrading bacterium. *Applied and Environmental Microbiology*, 69(9), 5198-5206.

Singh, B., Kaur, J. & Singh, K. (2012) Biodegradation of malathion by *Brevibacillus* sp. strain KB2 and *Bacillus cereus* strain PU. *World Journal of Microbiology and Biotechnology*, 28, 1133–1141 https://doi.org/10.1007/s11274-011-0916-y.

Singh, B., Kaur, J. & Singh, K. (2012) Transformation of malathion by *Lysinibacillus* sp. isolated from soil. *Biotechnology Letters*, 34, 863–867. https://doi.org/10.1007/s10529-011-0837-8.

Singh, D. P., Khattar, J. I. S., Nadda, J., Singh, Y., Garg, A., Kaur, N. & Gulati, A. (2011). Chlorpyrifos degradation by the cyanobacterium *Synechocystis* sp. strain PUPCCC 64. *Environmental Science and Pollution Research*, 18(8), 1351-1359.

Sivaperumal, P., Kamala, K., & Rajaram, R. (2017). Bioremediation of industrial waste through enzyme producing marine microorganisms. *In Advances in Food and Nutrition Research*, 80, 165-179.

Sun, A., Liu, J., Shi, X., Li, D., Chen, J. & Tang, D. (2014). Marine bacterium strain screening and pyrethroid insecticide-degrading efficiency analysis. *Chinese Journal of Oceanology and Limnology*, 32(5), 1029-1035.

Vidali, M. (2001). Bioremediation. an overview. *Pure and Applied chemistry*, 73(7), 1163-1172.

Xu, G., Li, Y., Zheng, W., Peng, X., Li, W. & Yan, Y. (2007). Mineralization of chlorpyrifos by co-culture of *Serratia* and *Trichosporon* spp. *Biotechnology Letters*, 29(10), 1469-1473.

Xu, G., Zheng, W., Li, Y., Wang, S., Zhang, J. & Yan, Y. (2008). Biodegradation of chlorpyrifos and 3, 5, 6-trichloro-2-pyridinol by a newly isolated *Paracoccus* sp. strain TRP. *International Biodeterioration & Biodegradation*, 62(1), 51-56.

Yadav, I. C. & Devi, N. L. (2017). Pesticides classification and its impact on human and environment. *Environmental Science and Engineering*, *6*, 140-158.

Yadav, M., Srivastva, N., Singh, R. S., Upadhyay, S. N. & Dubey, S. K. (2014). Biodegradation of chlorpyrifos by *Pseudomonas* sp. in a continuous packed bed bioreactor. *Bioresource Technology*, 165, 265-269.

Zhang, S., Qiu, C. B., Zhou, Y., Jin, Z. P. & Yang, H. (2011). Bioaccumulation and degradation of pesticide fluroxypyr are associated with toxic tolerance in green alga *Chlamydomonas reinhardtii. Ecotoxicology*, 20(2), 337-347.

Zhang, W., Lin, Z., Pang, S., Bhatt, P. & Chen, S. (2020). Insights into the biodegradation of lindane (γ-hexachlorocyclohexane) using a microbial system. *Frontiers in Microbiology*, 11, 522.

Chapter 6

The Impact of Microorganisms on Anti-Fouling in Marine Industries

Nurul Najihah Rawi[1]
Nor Izzati Abd Rahman[1]
Mujahidah Mohd Ramzi[1]
Fazilah Ariffin[2]
Kesaven Bhubalan[1,2]
Jasnizat Saidin[1]
Nor Atikah Mohamed Zin[1]
Julius Yong Fu Siong[1]
Kamariah Bakar[1]
and Noraznawati Ismail[1,*]

[1]Institute of Climate Adaptation and Marine Biotechnology, Universiti Malaysia Terengganu, Kuala Nerus, Terengganu, Malaysia
[2]Faculty of Science and Marine Environment, Universiti Malaysia Terengganu, Kuala Nerus, Terengganu, Malaysia

Abstract

The colonization of fouling organisms on submerged surfaces in the marine environment is known as biofouling. This issue has drawn world attention as it causes serious problems for marine industries, resulting in large amounts of economic loss. Many chemical coatings have been used

---

* Corresponding Author's Email: noraznawati@umt.edu.my.

In: Climate Change and Environmental Perturbations
Editors: T. Min Pau, M. Abd. Ghaffar, M. Danish-Daniel Abdullah et al.
ISBN: 979-8-89113-271-9

to treat biofouling especially tributyltin (TBT). However, TBT becomes harmful to marine ecosystems as it also kills untargeted organisms. Taking this into consideration, the application of natural antifoulants is an alternative way to eradicate biofoulers without harming other marine organisms. Microorganisms associated with marine organisms are potent sources of antifoulants as they have the ability to inhibit the growth of other organisms. This chapter reviews the bacterial species that have compounds that could potentially inhibit the growth of foulers based on the publications from 2011 to 2021.

Keywords: marine bacteria, anti-fouling, natural products, bio-film, marine invertebrates

Introduction

Biofouling is defined as unpleasant complex layers consisting of marine organisms that form on any surface in contact with seawater (Dahms & Dobretsov, 2017). These biofouling structures are commonly formed through successive colonization processes of marine communities, ranging from microfoulers to macrofoulers (Ramasubburayan et al., 2014). Submerged materials are exposed to the settlement of thousands of marine micro-organisms, such as bacteria, fungi, phytoplankton, and diatoms, which are classified as microfoulers (Figure 1(i)). Meanwhile, macrofoulers consist of organisms such as barnacles, limpets, oysters, and algae (Figure 1(ii), (iii), (iv), and (v)). About 4,000 different species have been reported as fouling organisms, with more than 100 of them having the potential to cause huge damage to surfaces immersed in seawater (Jorge et al., 2012).

Figure 1. i) Example of a marine diatom (*Amphora sp.*) which acts as a microfouler on submerged surfaces; ii) fouling organisms isolated and identified from submerged surfaces, Barnacle; iii) limpet; iv) oyster; and v) algae.

The fouling process usually occurs in four distinct phases (Figure 2). First, the adsorption of dissolved organic molecules, such as proteins, glycoproteins, carbohydrates, and polysaccharides, takes place as soon as the surface is submerged in seawater. These organic molecules form a conditioning biofilm within a few minutes, which becomes susceptible to fouling organisms (Xie et al., 2019). Next, microbial populations of bacteria and diatoms begin to adhere to the biofilm, forming the primary colonizers. The process of transforming into the biofilm layer takes several hours. The primary colonizers act as the dominant components before the adherence of macrofouler organisms (Marhaeni et al., 2011). Once the formation of the biofilm is completed, the surface becomes environmentally suitable for the attachment of macrofoulers. In the third phase, soft macrofoulers (e.g., algae and marine invertebrate larvae) attach within a few days. The presence of larvae functions as the settlement indicator in the overall formation of biofouling (Cho et al., 2012). In the final phase, hard microfoulers (e.g., barnacles, mussels, molluscs, and macroalgae) start to grow and inhabit the surfaces. This process takes a period of several months (Chen et al., 2021).

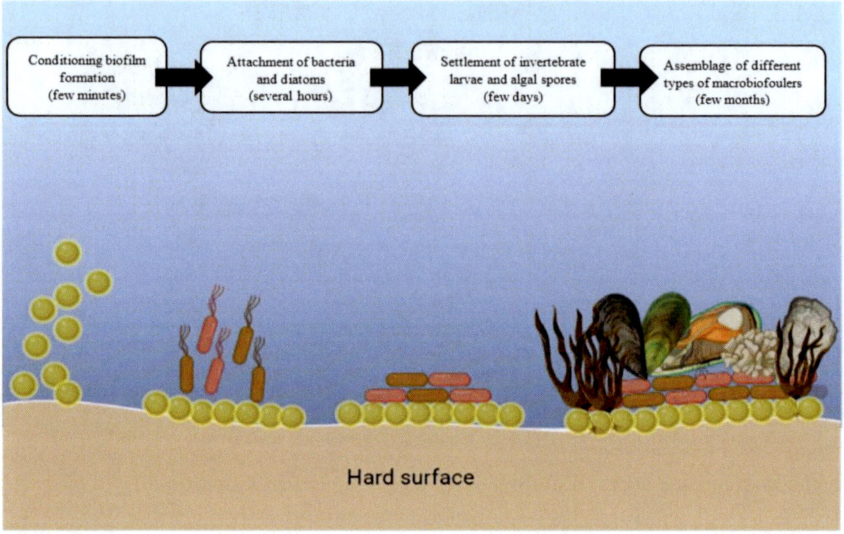

Figure 2. Formation of biofouling on a submerged surface.

Biofouling leads to numerous major problems in many important industries, especially for marine-based sectors such as shipping, power plants, aquaculture, and mariculture (Nalini et al., 2019). The impacts can be clearly observed in terms of massive economic losses, as well as from environmental aspects and safety concerns (Chen et al., 2021). For example, the assemblage of biofoulers on the underwater surfaces of a ship's hull elevates the level of frictional resistance, automatically causing an increase in total fuel usage of the ship (Hakim et al., 2019). At the same time, this condition consequently leads to an increase in the volume of harmful greenhouse gases released into the atmosphere. Furthermore, the fouling organisms that are attached to the ship structures will be carried along the ship's transportation path, therefore potentially introducing invasive species into non-native countries, eventually altering the balance of local ecosystems.

Furthermore, biofouling affects the submerged structures by increasing the rate of corrosion and accelerating the damage to underwater industrial equipment and facilities. This condition is of serious concern to the public, due to the rising safety issues. Biofouling that takes place in the inner part of pipeline systems reduces the performance and integrity of the facilities, especially when considering power plant infrastructure. In the aquaculture industry, the accumulation of biofoulers on the cage netting frame not only causes huge material losses, but these unwanted organisms also interfere with the process of nutrient exchange in that particular environment (Xie et al., 2019). Therefore, a significant amount must be spent on maintenance and periodic cleaning (Ramasubburayan et al., 2017a). Considering all of these concerns, this chapter summarizes promising anti-fouling compounds originating from marine microbes associated with marine invertebrates.

Commercial Anti-Foulants in the Market

Biofouling problems must be immediately addressed, in order to prevent the escalation of future problems faced by marine ecosystems and/or economically important industries. The process of controlling fouling on a surface is known as anti-fouling. The techniques used in this process include cleaning, biocide usage, and coatings (Archana & Sundaramoorthy, 2019). Other current approaches used to address biofouling problems include electrolyzing water, freshwater flushing, laser radiation, and photocatalysis (Chen et al., 2021).

The most ubiquitous technique used to prevent fouling is chemical treatment using biocides and coatings on the submerged surface. Previously, anti-fouling coatings based on organotin compounds such as tributyltin (TBT) and triphenyltin were regularly used (Zulkifli et al., 2019). According to Archana & Sundaramoorthy (2019), TBT has been continuously used since the 1970s as the most effective chemical to inhibit biofoulers, with high monetary returns to the relevant industrial players. However, due to the intensive use of anti-fouling coatings containing TBT on the outer surfaces of ships, these components may be released in harmful amounts and accumulate in marine environments. The influx of these toxic compounds (Soroldoni et al., 2018) could reach more than five tons annually on a global scale (Karlsson et al., 2010). This leads to water pollution and poses a threat to marine ecosystems, due to the high toxicity of the chemical. In fact, the application of TBT has been banned in the shipping industry by the International Maritime Organization (IMO), effective in September 2008 (Darya et al., 2020), although it is still widely used in other industries.

Supplementary booster biocides, such as Irganol, Diuron, SeaNine 211, and dichlofluanid have more recently been introduced onto the market. Though these biocides are more effective against biofoulers, they have also been reported as posing substantial threats to marine ecosystems, as they also unfavorably affect non-target marine organisms (Ramasubburayan et al., 2014; Chen & Qian, 2017). These toxic biocides are still in production, even though non-toxic anti-fouling methods are now available in the market, due to the higher production costs and lower effectiveness of the latter (Trepos et al., 2014; Dahms & Dobretsov, 2017).

Natural Anti-Foulants Isolated from Marine Bacteria

The application of natural anti-foulants provides an alternative means to prevent biofoulers without harming non-target marine organisms in the marine environment. A novel approach using marine natural products is required to produce eco-friendly and environmentally friendly coatings (Nalini et al., 2019). The development and investment into research on non-toxic coatings derived from natural compounds of marine organisms have been prompted due to the awareness regarding toxic anti-foulants. Promising sources for natural anti-foulants may be isolated from marine and terrestrial invertebrates, seaweeds and aquatic plants, marine organisms, and other sources, due to their specificity compared to heavy metal-based coatings (Ma et al., 2017). Several studies have shown that sponge compounds contain efficient anti-fouling agents; for example, guanidine alkaloids derived from the sponge Crambe crambe have been shown to possess anti-fouling, anti-microbial, and anti-predation properties (Acevedo et al., 2013).

Anti-Fouling Compounds from Marine Bacteria

Micro-organisms associated with marine organisms are potent sources for this purpose, as they have the ability to inhibit the growth of other organisms. These micro-organisms secrete bioactive compounds in order to protect themselves and survive in the marine environment. Marine bacteria have been shown to produce a high percentage of bioactive compounds, compared to other microbes (Viju et al., 2015). Many marine bacteria exhibit inhibitory activities against biofouling organisms through these bioactive compounds, also known as secondary metabolites. The functional groups of compounds such as steroids, phenolics, terpenoids, carotenoids, peptides, furanones, lactones, and alkaloids have been reported to show anti-fouling activities (Feng et al., 2018). A list of bacterial species producing bioactive compounds against biofouling is provided in Table 1. Based on a previous study by Kharchenko et al., (2020), a bacterial extract could substitute the cuprous biocide content by 50% when incorporated into a water-based paint resin. It showed the same performance in anti-fouling activity by preventing the formation of biofilm caused by the settlement of animal organisms.

Based on the list of bacterial species studied from 2011 to 2021, it can be observed that *Bacillus* sp. is the most common producer of bioactive compounds for anti-fouling purposes, as shown in Figure 3. Additionally, 46% of the previous works described that *Bacillus* sp. was likely to inhibit or reduce the growth of biofilm-forming bacteria. Common biofilm-forming bacteria in the marine environment consists of members of the genera *Vibrio*, *Pseudomonas,* and *Shewanella*, where the members of genus *Vibrio* were typically dominant in the biofouling communities studied by Waturangi et al., (2017).

Table 1. Anti-fouling bioactive compounds derived from various species of marine bacteria

Marine bacteria strains	Host	Active compounds	Anti-biofouling activities	Reference
B. subtilis MUT:M15	Cuttlefish	Phthalate derivatives	Reduced the settlement of microfouling and macrofouling organisms.	(Viju et al., 2019)
B. cereus	Sponges	Carbonyl groups	Inhibited the activity of biofilm-forming bacteria, *Alteromonas* sp. and *Pseudomonas* sp.	(Satheesh et al., 2012)
Actinobacteria species	Coral	-	Inhibited the biofilm formation of *Vibrio neocaledonicus.*	(Waturangi et al., 2017)
Streptomyces sampsonii PM33	Mangrove	Taxifolin	Inhibited algal spores and attachment of molluscs.	(Gopikrishnan et al., 2019)
S. violaceoruber	Seaweed	Furanone derivatives	Presented anti-biofouling activities against fouling organisms (e.g., zoospores, diatoms, and mussels).	(Hong & Cho, 2013)
B. subtilis subsp.	Oyster	Alcohol, alkenes, ethers, ketones, carboxylic acid and amines	Reduced biofilm formation antagonistically.	(Viju et al., 2018)

Table 1. (Continued)

Marine bacteria strains	Host	Active compounds	Anti-biofouling activities	Reference
B. licheniformis *B. subtilis*	Sponge	-	Active against marine bacterial strains in a broad spectrum.	(Aguila-Ramírez et al., 2014)
Vibrio alginolyticus	Sea anemone	Indole	Presented anti-biofouling activities against mussels, barnacles, and marine bacteria.	(Wang et al., 2017)
Bacillus sp.	Sponge	-	Inhibited diatom settlement.	(Jin et al., 2014)
Bacillus sp. *Virgibacillus* sp.	Seagrass	-	Inhibited the growth of biofilm-forming bacteria in a broad spectrum.	(Marhaeni et al., 2011)
Pseudomonas aeruginosa	Gastropod	Alcohol, alkanes, alkynes, amines, carboxylic acid and ethers	Inhibited the growth of biofilm-forming bacteria.	(Viju et al., 2016)
Bacillus sp. ICN-SS01	Seaweed	Exopolysaccharides	Reduced the adhesion of biofilm-forming bacteria.	(Rajasree et al., 2012)
Exiguobacterium sp.	Polychaete	-	Inhibited the adhesion of biofilm-forming bacteria.	(Rajasree et al., 2014)
P. taiwanensis	Seaweed	Alcohol, alkenes, amines, esters and carboxylic acid	Inhibited the growth of biofilm-forming bacteria.	(Viju et al., 2014)
P. putida	Octopus	-	Reduced the formation of bacterial biofilm.	(Viju et al., 2015)

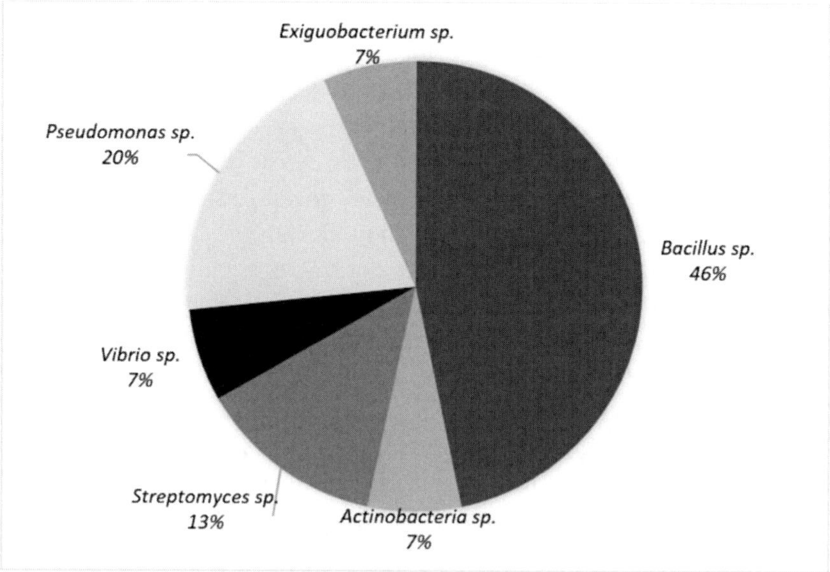

Figure 3. Bacterial strains reported as having anti-fouling activities from 2011 to 2021.

Members of genus *Bacillus* produce numerous secondary metabolites with many biological activities, including anti-fungal, anti-microbial, and cytotoxic effects (Jin et al., 2014). *Bacillus* species that are associated with different host marine organisms have been shown to prevent the formation of biofilm-forming bacteria, the settlement of barnacles (*Balanus amphitrite*), and algal spores (of *Ulva lactuca*). In a study by Satheesh et al. (2012), *Bacillus* sp. showed biofilm inhibitory activity against bacteria and microalgae. *Bacillus* spp. is classified as Gram-positive bacteria which produces secondary metabolites with anti-fouling activity. Gram-positive bacteria are well-known as producers of biologically active secondary metabolites for many purposes. Studies have indicated that more than 3% of Gram-positive bacteria contain genes for the metabolic synthesis of active compounds, such as a non-ribosomal peptide synthetase (Aguila-Ramírez et al., 2014).

Anti-Fouling Compounds from Marine Bacteria Associated with Marine Invertebrates

Sponges

Sponges are marine invertebrates that have developed distinctive ways to combat their predators and competitors; in particular, they are able to produce specific compounds against biofilm-forming micro-organisms (Selvin et al., 2010). Sponges are free from biofoulers through chemical defenses brought about by their extensive microbial associations. Many natural compounds isolated from various species of sponges have been verified as being microbially produced (Jin et al., 2014).

Sayem et al. (2011) have reported that *B. licheniformis*, a sponge-associated bacteria species from the marine sponge *Spongia officinalis*, can inhibit the formation of biofilms without disrupting the growth of the tested biofilm-forming bacteria *Escherichia coli* PHL628 and *Pseudomonas fluorescens*. The active compound in *B. licheniformis* cell-free supernatant was found to have a functional group of polysaccharides. This polysaccharide is composed of ∝-D-galactopyranosyl-(1→2)-glycerol-phosphate monomeric units, as shown in Figure 4. The authors investigated the ability of this compound to inhibit biofilm formation by two bacterial species: *E. coli* PHL628 and *P. fluorescens*. The anti-biofilm activity directly increased with the concentration of the bacterial supernatant, and the compound could inhibit biofilm formation for both species, with slightly higher inhibition for *E. coli* PHL628. The biofilm inhibitions were comparable, at 89% and 80% of the biofilm formation by *E. coli* PHL628 and *P. fluorescens*, respectively.

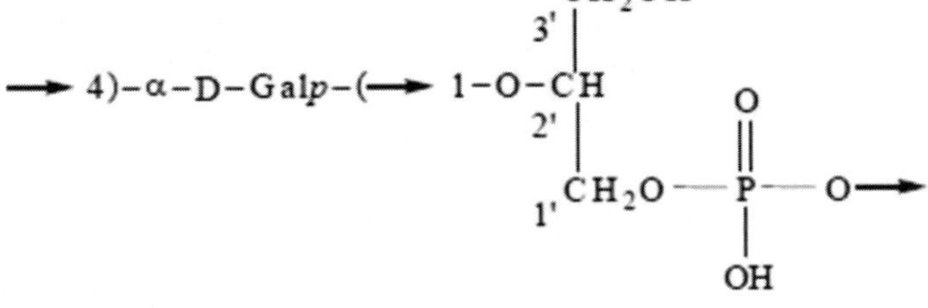

Figure 4. The repeating unit of polysaccharide obtained from *B. licheniformis* has anti-biofilm activity (Sayem et al., 2011).

This polysaccharide can affect biofilm formation for both Gram-positive and -negative bacteria, indicating that the compounds derived from *B. licheniformis* have broad-spectrum anti-biofilm activity, which helps them to survive on the host surface. The findings of Jiang et al., (2011) suggested that this polysaccharide may interfere with cell–surface and cell–cell adherence, due to the anti-adherence properties of the compound. Furthermore, the compound produced by *B. licheniformis* was reported to have a high content of phosphate groups, and it was proposed that the electronegative properties of this compound can regulate the reduction (or complete inhibition) of cell–surface or cell–cell interactions.

Seaweed

Bacteria associated with seaweed play an important role in helping to protect the seaweed from biofouling, through the secretion of secondary metabolites that act as anti-fouling compounds. The seaweed and their epibiotic bacteria are connected by a mutual relationship, in which both of the organisms obtain benefits from each other. Seaweed may provide nutrients for the surrounding ecosystem, while the bacteria protect the host seaweed from fouling organisms. Some marine invertebrates or seaweeds might inhibit the fouling process on their surface using both chemical substances and the help of surface-living bacteria (Rajasree et al., 2012).

The study of Hong and Cho (2013) revealed that the seaweed-associated bacteria *S. violaceoruber* isolated from *Undaria pinnafida* secreted two active compounds for anti-fouling activities, which were identified as furanone derivatives: (a) omF1, 3-octa-10,30-dienyl-4-methylfuran-2(5H)-one and (b) omF2, 3-octa-10-enyl-4-methylfuran-2(5H)-one (Figure 5). The LC_{50} and EC_{50} values, as well as the therapeutic ratio (LC_{50}/EC_{50}), were calculated. The therapeutic ratio can be used to evaluate the efficiency of anti-fouling compounds, in terms of their toxicity. Those with a ratio greater than 15 can be considered effective and non-toxic anti-foulants (Fusetani, 2011). The EC_{50} and LC_{50} values for compound (a) ranged from 0.03–0.1 μg/ml and 4.7–9.2 μg/ml, while those of compound (b) were in the range of 0.02–0.1 μg/ml and 4.2–7.5 μg/ml, respectively. The therapeutic ratio for both compounds was greater than 92 for three fouling organisms; namely, the mussel *Mytilus edulis* (representing hard biofoulers), the alga *U. pertusa,* and the diatom *N. annexa* (representing soft biofoulers).

Both of these compounds belong to the furanone group, members of which have been reported to exhibit many beneficial biological activities. Furanones are long-chain fatty acids that have been reported to reduce the thickness of the biofilm of *Staphylococcus epidermidis,* which was identified as pentadecanal (Casillo et al., 2017). Thomann et al. (2016) have also reported that furanone derivatives (i.e., 2-sulfonylpyrimidines) significantly inhibited the biofilm of *P. aeruginosa.* In short, two furanone derivatives isolated from the surface of seaweeds have shown significant anti-fouling properties. The compounds produced by the epibiotic bacteria of seaweeds may effectively protect the host against biofoulers.

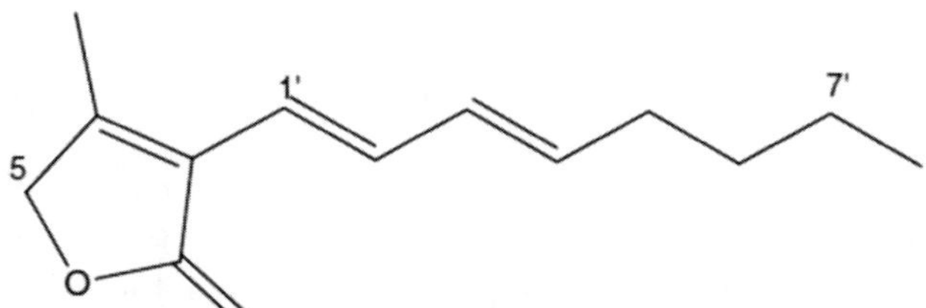

omF1

(a)

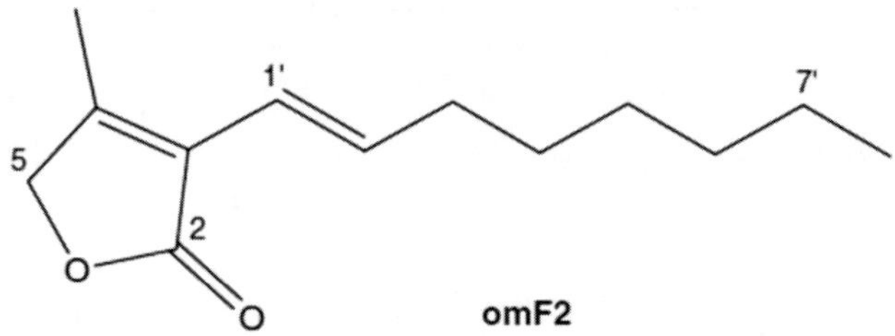

omF2

(b)

Figure 5. Structures of compounds isolated from culture extract of the bacterium *S. violaceoruber*: (a) 3-octa-10,30-dienyl-4-methylfuran-2(5H)-one; and (b) 3-octa-10-enyl-4-methylfuran-2(5H)-one (Hong & Cho, 2013).

Sea Anemones

Among all the marine fouling organisms, mussels are the most problematic fouling species, which can cause severe technical and economic losses (Callow & Callow, 2011; Schultz et al., 2011). They cause problems by producing silky strong byssal threads (fibres) to attach themselves to submerged solid surfaces. The accumulation of mussels can cause economic problems, as they can clog the water intake structures of piping systems, leading to decreased pumping function. Eventually, this problem will highly impact the maintenance costs of the industries and companies involved (Hosler, 2011).

In their study, Wang et al. (2017) showed that three small-molecule compounds from marine bacteria associated with a sea anemone (*Haliplanella* sp.) were effective against several biofoulers, including green mussel (*P. viridis*), barnacles (*B. albicostatus*), and marine bacteria (*Pseudomonas* sp.). These three compounds—indole, 3-formylindole, and cyclo(Pro-Leu) (Figure 6)—were isolated from an ethyl acetate extract of the marine bacterium *V. alginolyticus* PE2. Compounds (a) and (b) were reported to combat the byssus production of mussels through non-toxic mechanisms, while compound (c) led to mussel mortality due to its high toxicity. Compounds (a) and (b) slightly differ in structure, according to the presence of a formyl group in compound (b). The formyl group at position 3 in compound (b) was reported to decrease the anti-fouling activity against *P. viridis* and *Pseudomonas* sp., while increasing the activity against *B. albicostatus.* All of the compounds were significantly active as anti-fouling agents against byssal production by mussels, settlement of larval barnacles, and formation of bacterial biofilm. Among the compounds tested, indole exhibited the highest anti-fouling activity against *P. viridis*, with an EC_{50} of 24.45 μg/ml, while 3-formylindole had the highest activity in inhibiting the larval settlement of *B. albicostatus,* with an EC_{50} of 0.43 μg/ml. Moreover, the third compound, cyclo(Pro-Leu), had the highest potency in inhibiting the bacterial settlement of *Pseudomonas* sp. TB4, with no significant differences. As the fouling is made up of various organisms, a compound that inhibits a particular fouling organism may not be effective against other fouling organisms. In a previous study, Wang et al. (2015) stated that indole can inhibit the larval settlement of the bryozoan *Bugula neritina* and the barnacle *B. amphitrite*, thus suggesting indole as a promising potential compound to control the settlement of mussels in the future.

Coral

The marine environment is an expanded storehouse of natural products that possess the ability to prevent biofilm formation, including bioactive compounds obtained from corals, bryozoans, sponges, and pyrosomida (Tello et al., 2012). Coral ecosystems have an immense diversity of potential compounds originating from corals, compared to planktonic ecosystems. An example of corals that can inhibit the formation of biofilms are the soft coral *Antillogorgia elisabethae*, the black coral *Antipathes dichotoma*, the octocoral *Eunicea knighti*, and the cave coral *Acropora millepora* (Diana et al., 2016). Coral symbiotic microbes have become one of the most interesting sources for the screening of anti-fouling compounds, due to their high biodiversity and secretion of varied secondary metabolites (Golberg et al., 2013). Based on a previous study by Alagely *et al.* (2011), *Vibrio* spp. associated with coral was found to have an inhibitory effect on the formation of biofilms.

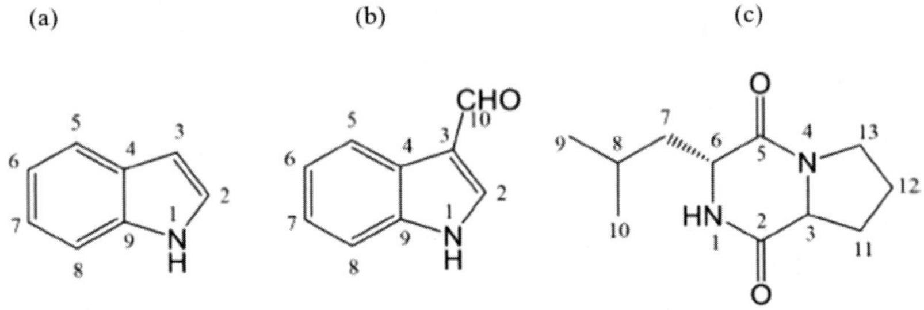

Figure 6. Chemical structures of anti-fouling compounds isolated from the marine bacterium *V. alginolyticus:* (a) Indole; (b) 3-formylindole; and (c) cyclo(Pro-Leu) (Wang et al., 2017).

Song et al., (2018) used an extract from a typical coral symbiotic bacterium, *V. alginolyticus* 12, which was found to possess anti-biofilm activity against the wild-type biofilm-forming isolate *P. aeruginosa* PA01. This potent compound was identified as a rhodamine isothiocyanate ($C_{29}H_{30}N_3O_3S$) analogue through combined GC-MS analysis and nuclear magnetic resonance (NMR). The chemical structure of this compound is shown in Figure 7. During the bacterial biofilm treatment using *P. aeruginosa* PA01 as the test strain, this compound was reported to reduce the biofilm density and thickness without affecting the viability of bacterial cells living in the biofilm. This property makes rhodamine isothiocyanate interesting and indicates its good potential to synergistically couple with other anti-fouling agents.

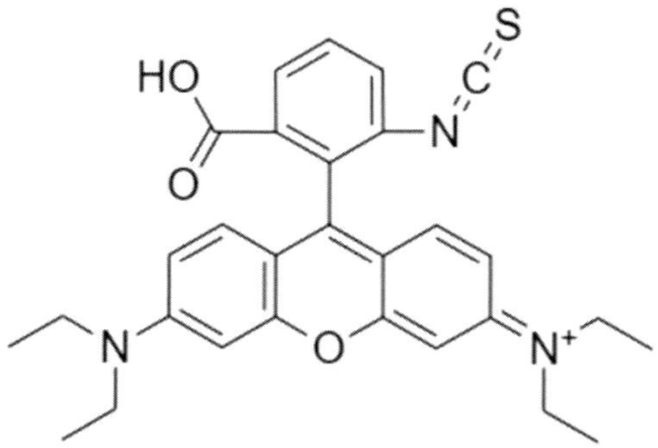

Figure 7. Structure of the compound isolated from *V. alginolyticus* 12, rhodamine isothiocyanate.

Advantages of Bacterially Produced Bioactive Metabolites

Most of the compounds with anti-fouling activities have been reported to be produced by marine organisms (Qian et al., 2015). Furthermore, marine bacteria are likely to produce distinctive substances that have not been identified in terrestrial organisms (Kharchenko et al., 2012). Many bioactive compounds with such activities have been isolated from marine invertebrates, such as sponges, coral, and gorgonians, as well as seaweed and algae. Based on the study of Satheesh et al., (2016), the metabolites which were believed to be produced by marine invertebrates for defensive purposes actually originate from their associated microbes. Hence, the supply issues of the compounds from those marine invertebrates are often hampered and prevent the commercial production of anti-fouling coatings (Marhaeni et al., 2011). Thus, intense research to identify the marine micro-organisms which produce bioactive metabolites to inhibit biofoulers is ongoing. On the other hand, micro-organisms have been proven to be an excellent source of such compounds without endangering non-target organisms in the marine ecosystem (Ramasubburayan et al., 2017b). The advantage of utilizing marine micro-organisms as a source of anti-fouling compounds is the capacity for large-scale production with a short growing time (Qian et al., 2009). These compounds can be produced in huge quantities in bioreactors by optimizing the culture conditions such as medium components, carbon and nitrogen sources, pH, and the temperature of the culture (Ramasubburayan et al., 2014).

Conclusion and Future Perspectives

This chapter successfully reviewed the bacterial species which produce compounds that may potentially inhibit the growth of biofoulers, based on the relevant literature published between 2011 and 2021. *Bacillus* sp. has been revealed to produce bioactive compounds that inhibit the growth of biofoulers. The production of such compounds can be increased through the optimization of culture conditions. The compounds must be non-toxic and exhibit suitable anti-fouling properties and must also be biodegradable in order to prevent their accumulation in the seabed and food web.

References

Acevedo, M. S., Puentes, C., Carreno, K., Leon, J. G., Stupak, M., Garcia, M., Perez, M. & Blustein, G. (2013), Antifouling paints based on marine natural products from Colombian Carribean. *International Biodeterioration and Biodegradation*, 83, 97-104. https://doi.org/10.1016/j.ibiod.2013.05.002.

Aguila-Ramirez, R. N., Hernandez-Guerrero, C. J., Gonzalez-Acosta, B., Id-Daoud, G., Hewitt, S., Pope, J. & Hellio, C. (2014). Antifouling activity of symbiotic bacteria from sponge *Aplysina gerardogreeni. International Biodeterioration and Biodegradation*, 90, 64-70. https://doi.org/10.1016/j.ibiod.2014.02.003.

Alagely, A., Krediet, C. J., Ritchie, K. B. & Teplitski, M. (2011). Signalling-mediated cross-talk modulates swarming and biofilm formation in a coral pathogen *Serratia marcescens. ISME Journal,* 5(10), 1609-1620. https://doi.org/10.1038/ismej.2011.45.

Archana, S. & Sundaramoorthy, B. (2019). Review on biofouling prevention using nanotechnology. *Journal of Entomology and Zoology Studies*, 7(4), 640-648.

Callow, J. A. & Callow, M. E. (2011). Friendly fouling-resistant marine coatings. *Nature Communications*. https://doi.org/10.1038/ncomms1251.

Casillo, A., Papa, R., Ricciardelli, A., Sannino, F., Ziaco, M., Tilotta, M., Selan, L., Marino, G., Corsaro, M. M., Tutino, M. L., Artini, M. & Parrilli, E. (2017). Anti-biofilm activity of a long-chain fatty aldehyde from Antarctic *Pseudoalteromonas haloplanktis* TAC125 against *Staphylococcus epidermis* biofilm. *Frontiers in Cellular and Infection Microbiology*, 7(FEB), 1-13. https://doi.org/10.3389/fcimb.2017.00046.

Chen, L. & Qian, P. Y. (2017). Review on molecular mechanisms of antifouling compounds: An update since 2012. *Marine Drugs*, 15(9), 1-20. https://doi.org/10.3390/md15090264.

Chen, L., Duan, Y., Cui, M., Huang, R., Su, R., Qi, W. & He, Z. (2021). Biomimetic surface coatings for marine antifouling: Natural antifoulants, synthetic polymers and surface microtopography. *Science of the Total Environment*, 766, 144469. https://doi.org/10.1016/j.scitotenv.2020.144469.

Cho, J. Y., Kang, J. Y., Hong, Y. K., Baek, H. H., Shin, H. W. & Kim, M. S. (2012). Isolation and structural determination of the antifouling diketopiperazines from marine-derived *Streptomyces praecox* 291-11. *Bioscience, Biotechnology and Biochemistry*, 76(6), 1116-1121. https://doi.org/10.1271/bbb.110943.

Dahms, H. U. & Dobretsov, S. (2017). Antifouling compounds from marine macroalgae. *Marine Drugs,* 15(9). https://doi.org/10.3390/md15090265.

Darya, M., Sajjadi, M. M., Yousefzadi, M., Sourinejad, I. & Zarei, M. (2020). Antifouling and antibacterial activities of bioactive extracts from different organs of the sea cucumber *Holuthuria leucospilota. Helgoland Marine Research*, 74(1). https://doi.org/10.1186/s10152-020-0536-8.

Feng, K., li, X. & Yu, L. (2018). Synthesis, antibacterial activity and application in the antifouling marine coatings of novel acylamino compounds containing gramine groups. *Progress in Organic Coatings*, 118(September 2017), 141-147. https://doi.org/10.1016/j.porgcoat.2017.10.027.

Fusetani, N. (2011). Antifouling marine natural products. *Natural Product Reports,* 28(2), 400-410. https://doi.org/10.1039/c0np00034e.

Golberg, K., Pavlov, V., Marks, R. S. & Kushmaro, A. (2013). Coral-associated bacteria, quorum sensing disrupters and the regulation of biofouling. *Biofouling*, 29(6), 669-682. https://doi.org/10.1080/08927014.2013.796939.

Gopikrishnan, V., Radhakrishnan, M., Shanmugasundaram, T., Ramakodi, M. P. & Balagurunathan, R. (2019). Isolation, characterization and identification of antifouling metabolite from mangrove derived *Streptomyces sampsonii* PM33. *Scientific Reports*, 9(1), 1-10. https://doi.org/10.1038/s41598-019-49478-2.

Hakim, M. L., Nugroho, B., Nurrohman, M.N., Suastika, I. K. & Utama, I. K. A. P. (2019). Investigation of fuel consumption of an operating ship due to biofouling growth and quality of anti-fouling coating. *IOP Conference Series: Earth and Environmental Science*, 339(1). https://doi.org/10.1088/1755-1315/339/1/012037.

Hong, K. & Cho, J. Y. (2013) Effect of seaweed epibiotic bacterium *Streptomyces violaceoruber* SCH-09 on marine fouling organisms. *Fisheries Science*, 79(3), 469-475. https://doi.org/10.1007/s12562-013-0604-y.

Hosler, D. M. (2011). Early detection of dreissenid species: Zebra/Quagga mussels in water systems. *Aquatic Invasions,* 6(2), 217-222. https://doi.org/10.3391/ai.2011.6.2.10.

Jiang, P., Li, J., Han, F., Duan, G., Lu, X., Gu, Y. & Yu, W. (2011). Antibiofilm activity of an exopolysaccharide from marine bacterium *Vibrio* sp. QY101. *PLoS ONE*, 6(4), 1-11. https://doi.org/10.1371/journal.pone.0018514.

Jin, C., Xin, X., Yu, S., Qiu, J., Miao, L., Feng, K. & Zhou, X. (2014). Anti-diatom activity of marine bacteria associated with sponges from San Juan Island, Washington. *World Journal of Microbiology and Biotechnology,* 30(4), 1325-1334. https://doi.org/10.1007/s11274-013-1557-0.

Jorge, P., Lourenco, A. & Pereira, M. O. (2012). New trends in peptide-based anti-biofilm strategies: a review of recent achievements and bioinformatic approaches. *Biofouling*, 28(10), 1033-1061. https://doi.org/10.1080/08927014.2012.728210.

Karlsson J., Ytreberg, E. & Eklund, B. (2010). Toxicity of anti-fouling paints for use on ships and leisure boats to non-target organisms representing three trophic levels. *Environmental Pollution,* 15893), 681-687. https://doi.org/10.1016/j.envpol.2009.10.024.

Kharchenko, U., Beleneva, I. & Karpov, V. (2020). Antifouling and anticorrosion potential of *Pseudomonas piscicida* extract. *Journal of Coatings Technology and Research*, 17(4), 887-896. https://doi.org/10.1007/s11998-019-00308-0.

Ma, C., Zhang, W., Zhang, G. & Qian, P. Y. (2017). Environmentally friendly antifouling coatings based on biodegradable polymer and natural antifoulant. *ACS Sustainable Chemistry and Engineering,* 5(7), 6304-6309. https://doi.org/10.1021/acssuschemeng.7b01385.

Marhaeni, B., Radjasa, O. K., Khoeri, M. M., Sabdono, A., Bengen, D. G. & Sudoyo, H. (2011). Antifouling activity of bacterial symbionts of seagrasses against marine biofilm-forming bacteria. *Journal of Environmental Protection*, 2(9), 1245-1249. https://doi.org/10.4236/jep.2011.29143.

Nalini, S., Inbakandan, D., Venkatnarayanan, S., Mohammed Riyaz, S. U., Dheenan, P. S., Vinithkumar, N. V., Sriyutha Murthy, P., Parthasarathi, R. & Kirubagaran, R. (2019). PYRROLO isolated from marine sponge associated bacterium *Halobacillus kuroshimensis* SNSAB01 – Antifouling study based on molecular docking, diatom adhesion and musselbyssal thread inhibition. *Colloids and Surfaces B: Biointerfaces*, 173(May 2018), 9-17. https://doi.org/10.1016/j.colsurfb.2018.09.044.

Qian, P. Y., Li, Z., Xu, Y., Li, Y. & Fusetani, N. (2015). Mini-review: Marine natural products and their synthetic analogs as antifouling compounds: 2009–2014. *Biofouling*, 31(1), 101-122. https://doi.org/10.1080/08927014.2014.997226.

Qian, P. Y., Xu, Y. & Fusetani, N. (2009). Natural products as antifouling compounds: recent progress and future perspectives. *Biofouling,* 26(2), 223-234. https://doi.org/10.1080/08927010903470815.

Rajasree, V., Satheesh, S. & Prakash Vincent, S. G. (2012). Antifouling activity of a marine epibiotic bacterium from the seaweed *Sargassum wightii. Thalassas: An International Journal of Marine Sciences*, 28(2), 37-43.

Rajasree, V., Shankar, C. V. S., Satheesh, S. & Punitha, S. M. J. (2014). Biofilm inhibitory activity of extracellular polymeric substance produced by *Exiguobacterium* sp. associated with the polychaete *Platynereis dumerilii. Thalassas,* 30(2), 13-19.

Ramasubburayan, R., Prakash, S., Iyapparaj, P., Sumathi, S., Titus, S., Immanuel, G. & Palavesam, A. (2017). Isolation, screening and evaluation of antifouling activity of mangrove associated bacterium, *Bacillus subtilis* subsp. *subtilis* RG. *Proceedings of the National Academy of sciences India Section B – Biological Sciences*, 87(3), 1015-1024. https://doi.org/10.1007/s40011-015-0674-y.

Ramasubburayan, R., Titus, S., Kumar, P., Immanuel, G. & Palavesam, A. (2017). Antifouling activity of marine epibiotic bacterium *Bacillus flexus* APGI isolated from Kanyakumari Coast, Tamilnadu, India. *Indian Journal of Geo-Marine Sciences*, 46(7), 1396-1400.

Ramasubburayan, R., Titus, S., Verma, P. K., Immanuel, G. & Palavesam, A. (2014). Isolation, screening and optimization of culture conditions for enhanced antibacterial activity by a marine epibiotic bacterium *Bacillus flexus* APGI aginst fouling bacterial strains. *Journal of Pure and Applied Microbiology,* 8(4), 2909-2920.

Satheesh, S., Soniamby, A. R., Shankar, C. V. S. & Josephine Punitha, S. M. (2012). Antifouling activities of marine bacteria associated with sponge (*Sigmadocia* sp.).

Journal of Ocean University of China, 11(3), 354-360. https://doi.org/10.1007/s11802-012-1927-5.

Satheesh, S., Ba-Akdah, M. A. & Al-Sofyani, A. A. (2016). Natural antifouling compound production by microbes associated with marine microorganisms – A review. *Electronic Journal of Biotechnology,* 21, 26-35. https://doi.org/10.1016/j.ejbt.2016.02.002.

Sayem, S. M. A., Manzo, E., Ciavatta, L., Tramice, A., Cordone, A., Zanfardino, A., De Felice, M. & Varcamonti, M. (2011). Anti-biofilm activity of an exopolysaccharide from a sponge-associated strain of *Bacillus licheniformis*. *Microbial Cell Factories*, 10, 1-12. https://doi.org/10.1186/1475-2859-10-74.

Schultz, M. P., Bendick, J. A., Holm, E. R. & Hertel, W. M. (2011). Economic impact of biofouling on a naval surface ship. *Biofouling*, 27(1), 87-98. https://doi.org/10.1080/08927014.2010.542809.

Selvin, J., Ninawe, A. S., Seghal Kiran, G. & lipton, A. P. (2010). Sponge-microbial interactions: Ecological implications and bioprospecting avenues. *Critical Reviews in Microbiology*, 36(1), 82-90. https://doi.org/10.3109/10408410903397340.

Song, Y., Cai, Z. H., Lao, Y. M., Jin, H., Ying, K. Z., Lin, G. H. & Zhou, J. (2018). Antibiofilm activity substances derived from coral symbiotic bacterial extract inhibit biofouling by the model strain *Pseudomonas aeruginosa* PAO1. *Microbial Biotechnology*, 11(6), 1090-1105. https://doi.org/10.1111/1751-7915.13312.

Soroldoni, S., Castro, L. B., Abreu, F., Duarte, F. A., Choueri, R. B., Moller, O. O., Fillmann, G. & Pinho, G. L. L. (2018). Antifouling paint particles: Sources, occurrence, composition and dynamics. *Water research*, 137, 47-56. https://doi.org/10.1016/j.watres.2018.02.064.

Tello, E., Castellanos, L., Arevalo-Ferro, C. & Duque, C. (2012). Disruption in quorum-sensing systems and bacterial biofilm inhibition by cembranoid diterpenes isolated from the octocoral *Eunicea knighti. Journal of Natural Products,* 75(9), 1637-1642. https://doi.org/10.1021/np300313k.

Thomann, A., Brengel, C., Borger, C., Kail, D., Steinbach, A., Empting, M. & Hartmann, R. W. (2016). Structure-activity relationship of 2-sufonylpyrimidines as quorum-sensing inhibitors to tackle biofilm formation and eDNA release of *Pseudomonas aeruginosa. ChemMedChem,* 11(22), 2522-2533. https://doi.org/10.1002/cmdc.201600419.

Trepos, R., Cervin, G., Hellio, C., Pavia, H., Stensen, W., Stensvåg, K., Svendsen, J. S., Haug, T. & Svenson, J. (2014). Antifouling compounds from the sub-arctic ascidian *Synoicum pulmonaria*: synoxazolidinones A and C, pulmonarins A and B, and synthetic analogues. *Journal of Natural Products*. 77(9), 2105-2113. https://doi.org/10.1021/np5005032.

Viju, N., Anitha, A., Vini, S. S., Shankar, C. V. S., Satheesh, S. & Punitha, S. M. J. (2014). Antibiofilm activities of extracellular polymeric substances produced by bacterial symbionts of seaweeds. *Indian Journal of Geo-Marine Sciences*, 43(11), 2136-2146.

Viju, N., Satheesh, S. & Punitha, S. M. J. (2015). Antifouling activities of antagonistic marine bacterium *Pseudomonas putida* associated with an octopus. *Proceedings of the National Academy of Sciences India Section B – Biological Sciences*, 87(4), 1113-1124. https://doi.org/10.1007/s40011-015-0691-x.

Viju, N., Punitha, S. M. J. & Satheesh, S. (2019). Antifouling potential of symbiotic marine bacterium *Bacillus subtilis* mut:M15 associated with the cuttlefish *Sepia* sp. *Indian Journal of Geo-Marine Sciences*, 48(5), 685-697.

Viju, N., Punitha, S. M. J. & Satheesh, S. (2018). Antifouling properties of bacteria associated with marine oyster *Crassostrea* sp. *Thalassas,* 34(2), 471-482. https://doi.org/10.1007/s41208-018-0095-9.

Viju, N., Satheesh, S. & Punitha, S. M. J. (2016). Antibiofilm and antifouling activities of extracellular polymeric substances isolated from the bacteria associated with marine gastropod *Turbo* sp. *Oceanological and Hydrobiological Studies*, 45(1), 11-19. https://doi.org/10.1515/ohs-2016-0002.

Wang, K. L., Xu, Y., Lu, L., Li, Y., Han, Z., Zhang, J., Shao, C. L., Wang, C. Y. & Qian, P. Y. (2015). Low-toxicity diindol-3-ylmethanes as potent antifouling compounds. *Marine Biotechnology,* 17(5), 624-632. https://doi.org/10.1007/s10126-015-9656-6.

Wang, X., Huang, Y., Sheng, Y., Su, P., Qiu, Y., Ke, C. & Feng, D. (2017). Antifouling activity towards mussel by small-molecule compounds from a strain of *Vibrio alginolyticus* bacterium associated with sea anemone *Haliplanella* sp. *Journal of Microbiology and Biotechnology,* 27(3), 460-470. https://doi.org/10.4014/jmb.1607.07068.

Waturangi, D. E., Hariyanto, J. P., Lois, W., Hutagalung, R. A. & Hwang, J. K. (2017). Inhibition of marine biofouling by aquatic actinobacteria and coral-associated marine bacteria. *Malaysian Journal of Microbiology*, 13(2), 92-99.

Xie, Q., Pan, J., Ma, C. & Zhang, G. (2019). Dynamic surface antifouling: Mechanism and systems. *Soft Matter,* 15(6), 1087-1107. https://doi.org/10.1039/c8sm01853g.

Zulkifli, M. F. R., Radzi, N. M., Jusoh, S. M., Saidin, J. & Nik, W. M. N. W. (2019). Potential of cabbage extract (*Brassica oleracea*) as anti-fouling agent in alkyd undercoat for mild steel in seawater. *Malaysian Journal of Analytical Sciences*, 23(3), 451-461. https://doi.org/10.17576/mjas-2019-2303-09.

Conclusion

Natural systems around the world are being affected by regional climate changes, such as temperature increases, which are very likely to be the result of anthropogenic activities. At the most basic level of biodiversity, climate change is able to decrease the genetic diversity of populations due to directional selection and rapid migration which, in turn, may affect ecosystem functioning and resilience. Most biodiversity resources are consumed and exploited by humans; therefore, it is our primary responsibility to preserve and protect biodiversity to protect natural environments worldwide. The richness of species diversity, the ecosystem, the environment and the sustainable growth of life on Earth are all of utmost importance. Biological resources have vast potential to support our wellbeing, from providing food to medicinal products. Unfortunately, we are also paying the price for the unsustainable exploitation of these resources, with over-harvesting, intensive farming (including aquaculture), and burning of fossil fuels leading to climate changes. The negative impacts of climate change on our local marine aquatic organisms are already being felt, as illustrated here. Moving forward, we must work together and develop specific and structured programmes to manage ecosystems and communities, allowing them to adapt and become more resilient in the face of these challenges. Protecting and restoring healthy ecosystems may be one of the key ways to mitigate and adapt to climate change.

This book represents our small contribution to address some of the current biodiversity issues plaguing the environment we live in. The vast marine biodiversity supported by our marine ecosystem has inspired innovative research by the scientists at ICAMB. These chapters are a representation of our work at the Institute; however, by no means have we provided a comprehensive account of the overall research conducted. Last, but not least, we are looking forward to producing more scientific evidence that will eventually be beneficial to all stakeholders, especially Mother Nature.

Min Pau Tan and Siti Azizah Mohd Nor

About the Editors

Associate Professor Dr. Tan Min Pau was born in a Peranakan Baba Nyonya family in Muar, Johor, Malaysia, on 5th September 1984. She is the second child among three siblings, and grew up at Chukai, Kemaman, Terengganu and had her early education in Kemaman before pursuing a Bachelor of Applied Science (Biology) at Universiti Sains Malaysia (USM) in 2004. She completed her bachelor's degree majoring in aquatic biology and minoring in management in 2007. She then enrolled in the doctoral degree in USM in 2008, and was awarded the USM Fellowship in 2009. She was awarded a doctorate degree in Biotechnology (specializing in population genetics) in 2011. At present, she works as a lecturer and holds administrative post as the Deputy Director at the Institute of Climate Adaptation and Marine Biotechnology (ICAMB), Universiti Malaysia Terengganu (UMT). Her research interest is the conservation of genetics, which includes aspects of the disciplines of population genetics, molecular ecology, phylogeography, evolutionary systematics, and others. Personally, she loves to live a healthy lifestyle, pursue adventurous activities, and deeply respects Mother Nature.

Professor Dato' Dr. Mazlan Abd. Ghaffar was born in Kuala Terengganu, Terengganu on 18th April 1962. He obtained a Bachelor of Science from Universiti Putra Malaysia (Malaysia) in 1985 and a Master of Science from University of Newcastle (England) in 1988. In 2001, he was awarded the Doctor of Philosophy degree from University of Wales (United Kingdom), after passing his study on fish biology and environmental ecophysiology. He is currently the Vice Chancellor of Universiti Malaysia Terengganu (UMT), and also serves as a researcher and academic staff member at the Faculty of Science and Marine Environment, UMT. His research interests lie majorly in fish and fisheries sciences, specifically but not limited to climate change resiliency in estuarine fishes, bio-physiology and digestion in fishes, and dynamic energy budget in air breathing fishes. At present, he is leading a multi-institutional mega research project called *Ocean climate change: Potential risk, impact and adaptation towards marine and coastal ecosystem services in Malaysia.*

Associate Professor Ts. Dr. Muhd Danish-Daniel Abdullah was born in Johor Bahru, Johor on 26th November 1980. He started his higher education in Universiti Kebangsaan Malaysia in 2001, and earned his first degree majoring in marine biology with distinction in 2004. He was then granted a National Science Fellowship by the Ministry of Science, Technology and Innovation to pursue his doctoral degree. He was conferred a Ph.D degree in marine science (specializing in marine microbiology and biotechnology) in 2009. At present, he serves as a researcher and academic staff member at the Institute of Climate Adaptation and Marine Biotechnology (ICAMB), Universiti Malaysia Terengganu, and is also the Head of the Bioinformatics Program at ICAMB. His research interests include (but are not limited to) the diversity and functional role of organisms in aquatic environments. He believes that sincerity in all his doings is the key to a successful, yet peaceful and happy life.

Dr. Mohamad Nor Azra was born in Port Dickson, Negeri Sembilan on 6th May 1987. He completed all of his higher-level education at Universiti Malaysia Terengganu, from Diploma (2008) to Bachelor (2011), Master (2013), and PhD (2019) level. He continued in his desire for research with the appointment of a Postdoctoral Fellowship at the same university, after which he was granted an Inaugural International Fellowship, the Leadership for Climate-Resilient Fisheries Fellowship (LCRFF), which was fully funded by the Sustainable Ocean Alliance and the Environmental Defence Fund, USA. He also received a fully funded Visiting Researcher at Research Center for Marine and Land Bioindustry, Earth Sciences and Maritime Organization, National Research and Innovation Agency (BRIN), Indonesia. At present, he serves as a researcher and academic staff member at the Institute of Climate Adaptation and Marine Biotechnology (ICAMB), Universiti Malaysia Terengganu, and is also the Head of the Climate Change and Adaptation Program at ICAMB. His research areas include: 1) The effects of climate change on marine aquatic organisms, 2) invasive species affecting environmental sustainability and societal wellbeing, and 3) physiological and behavioural changes in echinoderms. He strongly believes the quote "Stop making yourself so easily available for those who don't prioritize you."

List of Contributors

Amiruddin Ahmad Faculty of Science and Marine Environment, Universiti Malaysia Terengganu

Fazilah Ariffin Faculty of Sciences and Marine Environment, Universiti Malaysia Terengganu

Jasnizat Saidin Institute of Marine Biotechnology, Universiti Malaysia Terengganu, Terengganu

Julius Yong Fu Siong Institute of Climate Adaptation and Marine Biotechnology, Universiti Malaysia Terengganu

Kamariah Bakar Institute of Climate Adaptation and Climate Adaptation and Marine Biotechnology, Universiti Malaysia Terengganu

Kesaven Bhubalan Faculty of Sciences and Marine Environment and Institute of Marine Biotechnology Universiti Malaysia Terengganu

Liew Hon Jung Higher Institution Centre of Excellence (HICoE), Institute of Tropical Aquaculture and Fisheries (AKUATROP), Universiti Malaysia Terengganu

Mazlan Abd Ghaffar Faculty of Science and Marine Environment AND Higher Institution Centre of Excellence (HICoE), Institute of Tropical Aquaculture and Fisheries (AKUATROP), Universiti Malaysia Terengganu

Min Pau Tan Institute of Climate Adaptation and Marine Biotechnology, Universiti Malaysia Terengganu

Mohamad Nor Azra Md Adib Institute of Climate Adaptation and Marine Biotechnology, Universiti Malaysia Terengganu

Mohd Ezhar Mohd Noor Department of Agrotechnology and Bio-Industry, Politeknik Jeli Kelantan

Mohd Iqbal Mohd Noor Faculty of Business Management, Universiti Teknologi MARA (UiTM) (Pahang), Institute for Biodiversity and Sustainable Development, Universiti Teknologi MARA (UiTM), Shah Alam

Muhd Danish-Daniel Abdullah Institute of Climate Adaptation and Marine Biotechnology, Universiti Malaysia Terengganu

Mujahidah Mohd Ramzi Institute of Climate Adaptation and Marine Biotechnology, Universiti Malaysia Terengganu

Nadiah Wan Rasdi Faculty of Fisheries and Food Sciences, Universiti Malaysia Terengganu

Nor Atikah Mohamed Zin Institute of Climate Adaptation and Marine Biotechnology, Universiti Malaysia Terengganu

Nor Izzati Abd Rahman Institute of Climate Adaptation and Marine Biotechnology, Universiti Malaysia Terengganu

Noraznawati Ismail Institute of Climate Adaptation and Marine Biotechnology, Universiti Malaysia Terengganu

Nur Shafinaz Abu Bakar School of Marine and Environmental Sciences, Universiti Malaysia Terengganu

Nurul Najihah Rawi Institute of Climate Adaptation and Marine Biotechnology, Universiti Malaysia Terengganu

Siti Azizah Mohd Nor Institute of Climate Adaptation, Marine Biotechnology, Universiti Malaysia Terengganu

Siti Zuraida Zulkipli Institute of Climate Adaptation and Marine Biotechnology, Faculty of Fisheries and Food Science, Universiti Malaysia Terengganu

Tan Chun Hong Faculty of Science and Marine Environment and Institute of Oceanography and Environment (INOS), Universiti Malaysia Terengganu

Wee Hin Boo Institute of Climate Change, Universiti Kebangsaan Malaysia

Wen Hui Zhang Institute of Climate Adaptation and Marine Biotechnology, Universiti Malaysia Terengganu

Wen Jye Mok Institute of Climate Adaptation and Marine Biotechnology, Universiti Malaysia Terengganu

Yeong Yik Sung Institute of Climate Adaptation and Marine Biotechnology, Universiti Malaysia Terengganu

Zainudin Bachok Institute of Oceanography and Environment (INOS), Universiti Malaysia Terengganu.

Index

S

T

W